THE SOLUTION TO POLLUTION

MasterMedia books are available at a discount with bulk purchase for educational, group, premium, or sales promotion use. For information, please write or call:

Special Sales Department
MasterMedia Limited
215 Park Avenue South
Suite 1601
New York, NY 10003
(212) 260-5600

THE
SOLUTION
TO
POLLUTION

**101 ▶ Things
You Can Do
To Clean Up
Your Environment**

Laurence Sombke

MasterMedia Limited
NEW YORK

Library of Congress Cataloging-in-Publication Data

Sombke, Laurence.
  The solution to pollution: 101 things you can do to clean up your environment / Laurence Sombke.
    p. cm.
  ISBN 0-942361-19-9
  1. Recycling (Waste, etc.) 2. House cleaning. I. Title.
TX324.S66 1990
648—dc20
                                                            89-78376
                                                            CIP

Designed by Stanley S. Drate/Folio Graphics Co. Inc.

Manufactured in the United States of America.

10 9 8 7 6 5 4 3 2

CONTENTS

PREFACE

Ever since the first Earth Day, twenty years ago in April 1970, a lot of people have talked about cleaning up the environment. But very few people have done much about it. Yes, we do have the Clean Air and Clean Water acts of Congress, and industry has been forced to clean up *its* act, but what have we done as individuals to clean up our own backyard?

Very little. How many of you recycle your newspapers, glass bottles, and tin cans? Very few. That's the problem. We as individuals create a lot of garbage, and we throw a lot of garbage away. And up until now we didn't know and we didn't care what happened to that trash.

The lack of places to put our garbage is forcing all of us to come up with solutions. We are forced to act, forced to do something. We have got to stop creating so much garbage, so much hazardous waste, so much air and water pollution, or the delicate ecological balance of this planet is going to break down and get worse.

But how? What can I do? I'm just one person. What difference does it make if I recycle? It makes all the difference in the world. And there are things you can do, positive things you can do to stop pollution, and you don't have to radically change your lifestyle to do them.

That's why I wrote *The Solution to Pollution*. To give you practical, step-by-step, sensible, low-cost/no-cost tips and techniques that you can follow to clean up your little corner of the world.

And I have tried to separate fact from fiction about garbage. It doesn't do the planet any good to get up on a soapbox, point our fingers at everybody else, and blame the other guy. You have to have the facts to make intelligent decisions about recycling, energy conservation, and handling household hazardous waste.

There are hundreds of things you can do to clean up the environment. Most of them are listed in this book. Nobody expects you to be a saint or to live a monastic life all for the sake of ecological purity. But even if you do one little thing like saving your newspapers for the Boy Scout paper drive, or keeping your automobile tires properly inflated, or using latex rather than oil-based paint, you can make a difference. Every little bit helps, especially yours.

THE

SOLUTION

TO

POLLUTION

THE GARBAGE CAN:
A Recycling Primer

The next time you drive by the local garbage dump and hold your nose, remember, you helped create that mound of mess. The birds circling overhead are feasting on garbage that could have come from your house.

The next time your local beach is closed or you are warned against eating fish from coastal waters, lakes, and rivers because of pollution, remember, some of that sludge could have begun its journey from your backyard.

The next time you read a story about the global warming trend and the greenhouse effect, or your eyes burn from noxious polluted air, remember that the most mundane things you do in your daily routine are factors in that fouled atmosphere.

The world is a closed-loop system. Whatever pollution we release into the environment stays in the environment. It doesn't just go away. The oil spill of the *Exxon Valdez* in Alaska, the wandering garbage barge of two summers ago, the smog in our big cities, and human and medical waste washing up on our beaches are good examples of that fact.

Back in the good old days, we could simply bag up our garbage, put it on the curb, and wave bye-bye. But those days are drawing to a close in many parts of the country as our landfills are getting full.

According to the U.S. Environmental Protection Agency in Washington, the United States produces 160 million tons of solid waste, better known as garbage, each year. That's 1,300 pounds per person per year. That's 3.6 pounds of garbage per person per day. Eighty percent of that garbage goes to landfills. One-third of those landfills will reach capacity in the next five years.

Simply put, we are creating too much garbage and we are running out of places to put it. No one wants a garbage dump in their own backyard. NIMBY, it's called: Not In My Backyard. What's the solution?

The EPA says recycling. It has set a goal of reducing waste by 25 percent through recycling as well as source reduction by the year 1992. We now recycle less than 10 percent. Ten states have mandatory recycling laws now on the books and in force. Twenty other states are considering mandatory recycling. Nearly 1,000 communities have some type of mandatory recycling, including New York City, the nation's largest producer of garbage.

In this chapter, you are going to learn everything you ever wanted to know about recycling. Specifically, we will take a look at:

1. The type of household garbage we produce. Whether it be plastic, paper, glass, or metals, much of it can be recycled.

2. How garbage can be recycled, including techniques, strategies, tools, and resources.

3. How to turn your house into a recycling center.
4. Once it is recycled, how it is reused by industry.
5. How to start a recycling center in your community.

We are going to address the issues of biodegradable plastic—does it work? and incineration, or just plain burning trash—is it a good idea?

Let's get started. The first thing we should do is analyze what is in our household garbage cans. A simple way to do this for your household is on the night before your trash is picked up by the sanitation trucks, open a bag or two. Look at what is in there.

If your house is like most other American houses, here is what you will find:

41 percent is paper, newspapers, cardboard, paper food packaging.
8.7 percent is tin cans, aluminum cans, and other metals.
8.2 percent is glass jars, beer and beverage bottles.
6.5 percent is plastic bottles, containers, and bags.
7.9 percent is food waste.

The rest of our trash is composed of yard waste (17.9 percent), which we will discuss in another chapter, and durable goods like couches and other things that are difficult to recycle.

Americans produce a lot of trash, more than any other group of people in the world. Look at this graph.

Pounds of Garbage Produced per Person per Day

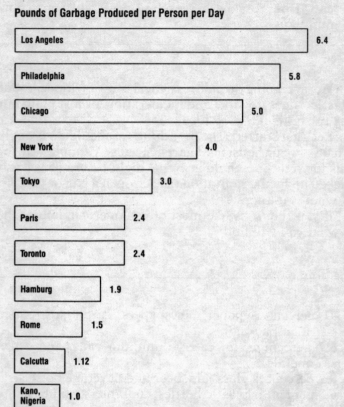

Los Angeles	6.4
Philadelphia	5.8
Chicago	5.0
New York	4.0
Tokyo	3.0
Paris	2.4
Toronto	2.4
Hamburg	1.9
Rome	1.5
Calcutta	1.12
Kano, Nigeria	1.0

In fact, the Environmental Defense Fund estimates that:

▶ We throw away enough glass bottles and jars to fill the 1,350-foot twin towers of New York's World Trade Center every two weeks.

▶ Every Sunday, more than 500,000 trees are used to produce the 88 percent of newspapers that are never recycled.

- ▶ Americans go through 2.5 million plastic bottles every hour.
- ▶ Every year, we dispose of 24 million tons of leaves and grass clippings that could be composted.
- ▶ We throw away enough iron and steel to continuously supply all the nation's automakers.
- ▶ American consumers and industry throw away enough aluminum to rebuild our entire commercial air fleet every three months.

That's a lot of garbage. Let's sort your garbage, see what's in there and see what and how we can recycle.

PAPER AND PAPERBOARD
(Also Known as Cardboard)

The earliest paper used in America was made from recycled rags. In 1690, the Rittenhouse Mill was opened near Philadelphia. It recycled cotton, linen, and used paper fibers into paper. As demand for paper grew, manufacturers turned to wood as a source of fiber to make paper, and fabric was phased out.

RECYCLING FACT: PAPER AND PAPERBOARD TAKE UP 41 PERCENT OF OUR LANDFILL SPACE.

RECYCLING FACT: EACH TON OF WASTEPAPER THAT IS RECYCLED SAVES 3.3 CUBIC YARDS OF LANDFILL SPACE.

The United States consumed 80.1 million tons of paper in 1986. We recycled 22 million of those tons; 13.5 million of those tons are in permanent use such as books, or flushed away as tissue; 42.5 million tons of paper were discarded and ended up in landfills.

Wastepaper that is recycled can be made into newspapers, writing and printing paper, paper bags, tissues, food boxes, cardboard displays, corrugated boxes, insulation in residential and commercial buildings, egg cartons, fruit trays, plant pots, cushioning material for packing crates and other things.

Unfortunately, all those cereal boxes and boxes of spaghetti, cat food, laundry soap, cookies, dress boxes, hat boxes, and waxed paper are just not recyclable. You will have to throw those away because the recyclers are not geared up to take them from consumers.

Newspapers and corrugated cardboard are the most important paper products for you to recycle in your home. Let's start recycling them now.

HOW TO RECYCLE NEWSPAPERS

1. Keep newspapers separate from all other types of paper.

2. Do not mix the shiny, colorful circulars in with newspapers even if they came delivered as part of your newspaper. This is particularly important to remember in separating the Sunday newspaper for recycling.

3. Do not mix mail from the post office in with the newspapers. The paper used in mail oftentimes is mixed with cellophane and other materials unfit for newspaper recycling.

4. Bundle your newspapers and stuff them in brown paper supermarket bags. Or tie them with string.

5. Companies that buy used newspapers for recycling are listed in the telephone commercial directory or Yellow Pages under Wastepaper.

6. Contact your local civic groups, such as the Boy Scouts or church groups, to see if a paper drive is under way. Or start one yourself by alerting your neighbors and getting them involved. Your municipal government public works office might also be interested in helping you get one started.

CORRUGATED BOXES

Corrugated boxes, better known as cardboard boxes, are primarily recycled by supermarkets, department stores, and other retail stores that receive their merchandise boxes packed in corrugated cardboard.

Common consumer corrugated boxes are refrigerator, air conditioner, and other appliance boxes, boxes for electronic equipment like TVs, stereos, computers, and typewriters. You may even get banana, apple, or orange boxes from the supermarket for some household task.

You, too, can recycle corrugated cardboard. Treat it just as you would newspaper. Separate it, form it into bundles, and tie it or bag it in brown paper bags. Recycle.

PLASTICS

Plastics make up the second largest pile of garbage that we toss into our trash cans and send to the dump. Plastic is only 7 percent of our garbage by weight, but 18 percent by volume. About half of that plastic is packaging. There are plastic soda bottles, plastic ketchup and salad oil bottles, plastic cups, plastic wrappings, egg cartons, milk jugs, detergents and cleaners, bread wrappers—the list goes on and on.

Complicating matters, food producers are catering to our fast-food-from-the-freezer-into-the-microwave mentality and creating a whole new line of plastic containers that sooner or later could end up in our garbage cans.

And, don't forget, we pack most of our garbage in huge thirty-gallon-capacity plastic garbage bags. A major part of our home economy comes wrapped in plastic, so it makes sense that such a large part of our garbage is comprised of plastic.

Here's a breakdown of our plastic garbage:

40.4 percent—garbage bags and film food wrappers.
28.4 percent—milk and detergent bottles.
9.5 percent—lids and heavy wrappers.
9.3 percent—clamshell food containers, foam plastic cups, cottage cheese and yogurt containers.

7

6.5 percent—soft drink bottles.
4.0 percent—cooking oil containers and other food wrappers.
1.9 percent—other.

You can see that the biggest portion of our plastic trash is garbage bags and film wrappers, which today are being recycled in only a few areas of the country. They can be reused, but most often not recycled.

The plastics industry is rather new. It didn't really get started until World War II, in the 1940s. The plastic one- and two-liter soft drink bottles we use now were introduced only in 1979. Admittedly, the plastics industry has been slow to consider recycling. But now they have several projects under way to recycle plastic, giving you, the household recycler, another opportunity to cut down on the items that end up in the garbage.

The biggest problem with recycling plastic is that there are several different types of plastic that we commonly use. Let's list the types of plastic, how they are used, how they can be recycled and used again.

PET—Polyethylene Terephthalate

One-, two-, and now three-liter soft drink bottles are made of PET. PET is also used in peanut butter, mouthwash, and mustard jars. It is also used as the plastic wrapping for "boil in bag" food pouches that can also be tossed in the microwave.

Seven hundred fifty million pounds of PET plastic bottles are consumed in the United States each year. About 170 million pounds, or 20 percent, are recycled.

Recycled PET is used as fibers for carpets, fiberfill, paint brushes, twine, rope, and scouring pads, industrial uses, and clear detergent bottles. Recycled PET is not now being used to make new soda bottles and other food containers because of fear that it might contain contaminates, although bottles of recycled PET will probably be lined with virgin PET in the near future, solving that problem.

PS—Polystyrene

Polystyrene (often incorrectly called Styrofoam, a Dow brand name for building insulation) has been around since 1938. Polystyrene is used to make cups, plates, and fast food containers, egg cartons, packing pellets, and clear clamshell containers used at fast food salad bars.

Until its banning by the food service industry in 1988, polystyrene foam containers were made with the use of fully halogenated chlorofluorocarbons (CFCs). When CFCs rise into the atmosphere, they break down, give off chlorine, and deplete the ozone layer.

Polystyrene can be recycled, but the facilities are just getting started. Recycled polystyrene is used to make office accessories, note pad holders, even waste cans. Rubbermaid has a set of waste cans made partially from recycled polystyrene.

PVC—Polyvinyl Chloride

Most PVC plastic is used by industry and not the homeowner. The exceptions are clear plastic cooking oil containers and some clear plastic meat wrappers, some peanut butter containers, some shampoo bottles, and some automobile products.

Although very little PVC is now being recycled, it can be used to make drainage and sewer pipe, bathtubs, floor tile, outdoor furniture, and other products.

HDPE—High-Density Polyethylene

Plastic milk bottles, butter tubs, soap, detergent, and bleach bottles, orange juice containers, heavy-duty trash bags and grocery bags, and other items are made with HDPE.

Ninety-three million pounds of HDPE were recycled in 1988. It can be recycled into lumber for boat piers, livestock pens, highway signs, flowerpots, toys, trash cans, kitchen drainboards, the base cup for soft drink bottles, detergent bottles, and many other things.

LDPE—Low-Density Polyethylene

LDPE is the flexible clear plastic that is used for wrappers for bananas, bread, and produce, dry cleaner bags, and many other uses. Because it is so soft and tearable, it is not currently being recycled on any large scale.

BIODEGRADABLE PLASTIC—DOES IT WORK?

There is a lot of talk these days about biodegradable plastic being the answer to our garbage crisis. Biodegradable plastic is plastic that will break down into small and smaller pieces when it is exposed to the atmosphere. Cornstarch is added to plastic to make it biodegradable. Photodegradable plastic is plastic that will break down into smaller pieces when it is exposed to sunlight.

But not everyone is convinced that bio- or photodegradable plastic is the right response to the garbage crisis. Many scientists and environmentalists—and plastics manufacturers themselves—are skeptical about bio- and photodegradable plastics. Many simply do not believe that it does, in fact, degrade. The plastics industry also is concerned about the negative impact that bio- or photodegradable plastic could have on recycling. They feel that recycling is the answer.

Scientists have found that hardly anything degrades in

a landfill because it is covered and not in contact with either light or air. They further note that modern landfills are better filled with inert items, those that do not degrade, because this way a landfill is solid and unchanging, so that no chemical gases can build up or hazardous wastes can leak out. Solid, dry, inert landfill can be covered and used for other uses on the surface level, such as parks, playgrounds, or commercial buildings.

PLASTICS AND WILDLIFE AND MARINE LIFE

Many of us have noticed garbage and medical waste washing up on our beaches each year. That's the good news, unfortunately, because that garbage can be picked up and disposed of properly. It is estimated that 14 billion pounds of trash are dumped into the ocean each year. Most of that illegal dumping is done by merchant vessels, although the U.S. Navy admitted to dumping off the coast of New Jersey in 1988. Merchant ships dump 1.5 billion plastic and glass containers into the ocean every day. The Coast Guard estimates that even the average boater or sports fisherman dumps over a pound of garbage into the ocean each time he goes out.

Much of that garbage is plastic. Plastic yokes for six-pack holders, plastic fishing line, plastic bags, plastic beverage containers, plastic motor oil containers, all get dumped into the ocean.

Besides the mess, the real tragedy is that marine animals and birds eat this plastic or get tangled up in it and die. The National Wildlife Federation says that plastic dumped in the ocean kills 10,000 animals each year. The National Marine Fisheries Service says 30,000 fur seals die each year because they get tangled up in plastic fishing nets left behind by trawlers.

Some people say that biodegradable and photodegradable plastic is the answer. But many people disagree. Many fear that people will be encouraged to litter in the oceans if they are convinced that their plastic is as degradable as bread.

The real solution is not to litter in the ocean. Don't throw anything in there. Put your litter in a bag and dispose of it back at the marina or at home. The Marine Plastic Pollution Research and Control Act, passed by Congress in 1987, makes it illegal for any vessel to dispose of plastics in U.S. waters up to the 200-mile limit.

GLASS RECYCLING

Unlike many other products, glass is 100 percent recyclable. Also, glass can be recycled over and over again. Today, glass containers manufactured in the United States are made with at least 25 percent recycled glass, called "cullet."

Glass is a very simple product. It is made with sand, limestone, soda ash, and recycled glass.

The kinds of glass the glass industry wants and needs to recycle are:

---

1. *Clear glass.* Peanut butter jars, jelly jars, beverage containers, all sorts. Two-thirds of recycled glass is clear.

RECYCLING TIP: PLACE YOUR CLEAR GLASS CONTAINERS, LIKE PEANUT BUTTER JARS, IN THE DISHWASHER TO CLEAN THEM BEFORE YOU RECYCLE.

The glass industry does not require your glass to be clean, or even that the labels be taken off, but food wastes can be a public health problem and create odors for your home.

2. *Brown glass.* One-quarter of recycled glass is

---

brown glass for beer bottles, apple juice, and prune juice bottles.

3. *Green glass.* Mostly wine bottles, with some beer and soft drinks, notably 7-Up.

Separating glass into the three colors is the best method of recycling glass. This is easy if you take your glass to a drop-off point where they can set up separate bins for the different colors. But, in the long run, it is more important to recycle glass than not, even if you mix the three colors of glass.

The glass industry likes to recycle glass because using recycled glass uses less energy to make new glass, meaning a cost savings for them.

RECYCLING ALUMINUM AND TIN CANS

Almost 14 million tons of metals, mainly tin, steel, and aluminum cans, go into the nation's trash cans each year. Only 8.7 percent of that is currently recycled. But metals, especially aluminum, are some of the easiest items to recycle and the ones that can produce the most cash for you or your civic group.

Aluminum

Aluminum is our favorite recyclable. In 1987, we recycled 50.5 percent of the aluminum cans we used. That's 36.6 billion aluminum cans recycled. Because aluminum is so valuable, can collectors earned $250 million for recycling aluminum in 1987.

The aluminum industry has always been in the forefront of recycling for economic reasons. Making aluminum from raw materials is very expensive because of the amount of electricity needed for the process. It only takes 5 percent as much electricity to make aluminum from recycled cans than if raw materials were used. In 1987 alone, the aluminum industry saved enough power by using recycled cans to supply the residential electric needs of New York City for six months.

There are now over 10,000 aluminum can recycling centers in the United States. Many are sponsored by companies like Reynolds Aluminum, some by beer breweries and distributors, and some by aggressive businesses.

Aluminum recycling centers or scrap dealers will take aluminum siding from your house, aluminum storm doors and windows, even aluminum pots and pans, as long as they are not contaminated with other metals. They will, although they don't like to, take aluminum foil and aluminum food trays. If they are clean, they are fine. But it is very difficult to get clean aluminum from these two sources.

Tin Can Recycling

Tin cans are really steel cans coated with tin. There are also some new steel cans coated with chromium, but for you, the recycler, it doesn't matter. All of your tin cans can be recycled. Industry will separate the tin, steel, and chromium after you recycle it.

In 1988, 29.1 billion food and beverage cans were produced and used. Approximately 15 percent of that was recycled. The steel industry says they could use 100 percent of all tin cans if we, the household recycler, could furnish them.

The reason is that both methods of manufacturing steel use scrap or recycled steel. The new regional mini-mills use 100 percent recycled steel to make nails, angle irons, and other heavy steel items. The potential is there.

Furthermore, using recycled steel saves steel producers 50 percent of their energy use and costs, savings that can make American steel makers more competitive and mean better prices for you.

Historically, paper wrappers and food waste had to be removed from tin cans to be recycled. But modern technology is not bothered by paper and food waste. Just toss the cans in the recycling bin. But if you are keeping the cans at home for a few days before you recycle, it is better from an odor point of view to rinse them.

INCINERATION VERSUS LANDFILL—TO BURN OR BURY

The recent big push for recycling in the United States is spurred by the closing down of our public landfills, better known as dumps. Since 1978, 70 percent of all landfill in the United States, that's 14,000 facilities, have shut their gates. The EPA predicts that by the year 2000, one-third of our existing landfills, plus the ones that are built in the interim, will close.

Here are some specifics about landfills:

▶ Five landfills have closed since 1987 in the Chicago metro area. The remaining twenty-nine cities will be full by 1995, according to Illinois officials.

▶ In 1992, Indiana will not be able to bury half of the state's trash because landfills will have dropped from 150 in 1980 to 53 in 1992.

▶ Pennsylvania's landfill capacity will be exhausted in five years.

▶ Connecticut has only two more years of landfill space left.

▶ New York will have no landfills in 1995.

Simply put, the garbage crisis in this country is a matter of too much garbage and too few places to put it. But we continue to create more garbage. In 1960, we produced only 2.5 pounds of garbage per person per day. In 1970, 2.8

pounds; in 1980, 3.25 pounds; in 1990, 3.6, and in the year 2000, the EPA projection is over 4 pounds of garbage per person per day.

There are three reasons why landfills are disappearing.

---

1. *Unsafe.* Many of our older landfills were built during a time when safety was not a high priority. Everything from paint, to leaking refrigerators, to car batteries was tossed in there. What happens now is that when rainwater settles through a dump site, leachate of hazardous and toxic chemicals leaks out of the site and into streams and groundwater. When state and federal EPAs find these dumps, they close them down and try to seal them.

2. *Full.* Because we are creating so much garbage, many of our dumps are simply filling up. The world's largest dump, the Fresh Kills site in Staten Island, New York City, will close in 1992.

3. *NIMBY.* Finally, when officials try to build new dumps, local citizens cry NIMBY, Not In My Backyard. Which is understandable. Nobody wants a garbage dump with huge trucks roaring in and out at all hours, birds circling overhead, and all the smells and noise of a dump so close to their property. Real estate values go down.

But not having landfill sites is causing major problems for business and government. In Columbia County, New York, where there is no open landfill, businesses are thinking of relocating to other areas and new business is staying away. The real estate market is stagnant.

Many governments are finding increasing portions of their budgets going to trash disposal because tipping fees

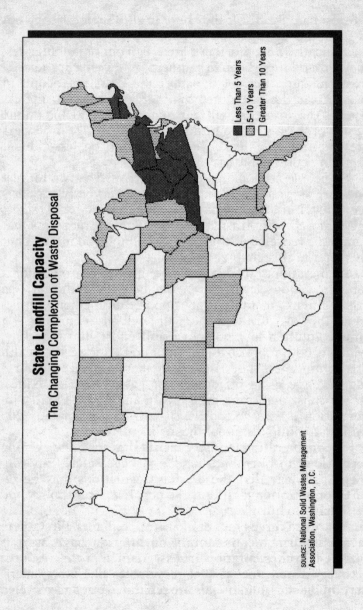

State Landfill Capacity
The Changing Complexion of Waste Disposal

Less Than 5 Years
5–10 Years
Greater Than 10 Years

SOURCE: National Solid Wastes Management
Association, Washington, D.C.

are going up locally. If they have to ship their garbage out of state, the fees are even higher.

Typically, when northern areas run out of landfill sites, they ship their garbage to southern, midwestern, and even western states. Besides the enormous costs, the people in those areas that are accepting the garbage are being exposed to hazards that will cause environmental and public health problems in the future. Well-off garbage producers are shunting their garbage problems off on less-affluent areas in poorer counties.

So people are looking for alternatives. The two biggest ones are recycling and incineration. Everybody supports recycling, but after the paper, certain plastics, and metals are removed from trash, there are simply some items that cannot be recycled at this time. (I will tell you more about this later in this chapter.)

Incineration, burning garbage, has been around for a long time and is getting a renewed boost from the garbage crisis. Environmentalists are opposed to incineration because they are afraid people will latch on to burning as a simple solution and not put a full effort into recycling, which is both more cost-effective and better for the health of a community. They also fear incineration because burning garbage gives off dangerous gases, nitrous oxides, which cause acid rain, even dioxin and furan. Finally, incineration is very expensive. It costs an estimated $200 million to build a state-of-the-art incinerator.

The incineration industry insists they have the existing technology to burn and handle garbage safely. Environmentalists fear that those safety regulations may be relaxed or forgotten if cities make burning public policy and push hard to burn garbage.

There are three types of incineration plants. First, there is mass burn, means simply burning garbage as it is dumped from sanitation trucks. Then there is resource recovery, which means that recyclables like newspapers, certain plastics, and metals are separated out and recycled before the garbage is burned. Finally, there is refuse-derived fuel, which means the garbage is burned and energy, either steam or electricity, is produced.

But even after incineration, there is ash. As much as a third of the volume of the original garbage ends up as ash after burning. At best, this ash is hazardous; at worst, it is toxic. It must be taken to a specially designed landfill that is monitored. It costs more to build these landfills, and it costs more to dump ash or landfill there.

It boils down to this: we are going to need more landfills, but let's make the most efficient use of them.

---

1. Let's recycle everything we possibly can. We are capable of recycling 64 percent of our garbage, which is much higher than our present 10 percent.

2. Let's use less. This is called "source reduction." I will tell you how later in this book.

3. As a last resort, let's incinerate what garbage we cannot recycle. Let's build incinerators that recycle again at the last minute. Make the incinerators environmentally sound and safe to operate.

4. Build modern and safe landfills.

HOW TO RECYCLE—TOOLS, TECHNIQUES, AND ACTIVITIES

Mandatory government-sponsored recycling is by far the most effective way to get the job done. Voluntary citizen-sponsored recycling is good, but it cannot reach the levels of mandatory action, like the 85 percent participation that Hamburg, New York, is reporting or the 80 percent in Islip, New York.

You can make mandatory recycling a reality by contacting your city and county officials, plus your elected state representative and state senator. Don't forget to write a letter to the governor of your state, also. Simply tell them that you support mandatory recycling, that you are a taxpayer and voter, and that you would like to participate in the decision-making process as a citizen.

But getting mandatory recycling into law can take several years. In the meantime, you can be very effective by working with civic groups and businesses in your area.

The first priority for starting a recycling campaign is to find a business that will accept your recycled materials. Look in the telephone book under Recyclers, Waste, Resource Recovery, or Scrap Yards. If that doesn't work, contact a local environmental group, or call the Environmental Defense Fund hot line at (800) CALL-EDF, or the National Recycling Coalition ([202] 659-6883), or the Environmental Protection Agency waste hot line ([800] 424-9346) for help. (The last chapter in this book is a national environmental and recycling resource information list. Check there, also.)

Ask the recycler what they will accept and in what form. Will they take newspapers? Do they have to be bundled in a certain way? What types of glass, aluminum, or tin cans will they take? What about plastic bottles? All of these are important because if you don't conform to the recycler's specifications, you will have to resort the trash and make a lot of extra work for yourself and your neighbors.

Next, you don't want to be the only person recycling, so try to find a civic, social, or religious group that is interested in participating.

Make recycling easy. Curbside programs, where the sanitation trucks come by your house on their regular rounds and pick up recyclables along with regular trash, is by far the most effective. Working directly with your trash haulers may be a good way to make this happen, but usually this takes a governmental mandate to get it done.

Without curbside pickup, you will have to deliver the

recyclables yourself. So, you are going to have to create drop-off spots, public sites in your neighborhood where people can bring their recyclables and place them in the proper containers. Supermarkets and shopping malls with huge parking areas are likely choices to approach. Remember, you will be responsible for keeping your drop-off sites neat and tidy. The recycler will then bring big trucks to the site to haul away the recyclables.

Regardless of whether you have curbside or drop-off-site recycling, you are going to have to sort your trash at home. Sorting your trash means you have to have separate containers for the different categories. You may get lucky, many recyclers want you to put newspapers in one pile and all other recyclables like tin cans, glass, and plastic in another one. They prefer to separate and sort the trash themselves. Chances are, though, that you will be doing the sorting.

Here are some home recycling tips:

1. Make sorting convenient and a part of your routine. Keep the receptacles easy to get to under the sink, on the back porch, in the basement. Recycling should be a natural, rather than cumbersome, part of your life.

2. Get several heavy-duty containers and label them for newspapers, glass, colored glass, plastics, or whatever your system is. Rubbermaid has developed a line of containers, made of recycled plastic, specifically designed for recycling. Hard plastic milk cases—the colorful ones that hold several gallons of milk—are also good recycling containers because they are compact, light, sturdy and you can stack them for home use or take them to the curbside.

3. Your containers need to be big enough to handle a week's worth of items, because that is about how often curbside or drop-site recycling takes place.

4. Your containers need to be sturdy but light enough for you to handle. Make sure they have heavy-duty handles so you can lift them.

5. You may want to have one main recycling bin in the kitchen that takes bottles and cans, which you separate into different containers on the porch or in the garage.

6. Do your recycling chores at a time of day that is convenient to you. Most cans, bottles, and jars need to be rinsed, so recycle while you are doing dishes.

7. Don't leave your recyclables out in the rain. It will add more weight to the containers and is a messy bother.

8. Don't leave recyclables in the backseat or trunk of the car. It's messy. Get organized and follow a system.

The Recycling Kitchen

If you have the time and inclination, there are ways you can redesign your kitchen to make it a smooth and efficient home recycling center. *The Smart Kitchen*, by David Goldbeck (Woodstock, NY: Ceres Press), has some very good ideas.

1. Place a recycling cart underneath your kitchen sink or in a nearby closet. These wire or hard plastic carts have three pull-out shelves. Fill one with newspapers, one with bottles and jars, and one with paper or plastic bags that you will reuse.

2. Build pull-out drawers into your lower kitchen cabinets. A large drawer, like an old-fashioned bread drawer, would be perfect for newspapers.

3. Build a large tilt-out drawer, with three compartments, one each for glass, metals, and plastic.

Make the compartments removable, so you can get the items to the recycling center.

4. Cut a hole in an unused portion of your counter-top and drop organic waste like potato peels and uneaten food into a compost bucket placed below. Install a sliding plastic or wooden cover for the hole so it can be opened and closed. Collect the organic waste and place it on a compost pile in your back-yard.

5. Install a slide-out stainless steel or hard plastic tray under your cutting area. When you have created a pile of organic waste—items like carrot peels—slide the tray out, scrape the waste in, and slide it back. At the end of the day, take the tray out to the compost pile. (Specifics on how to build a home compost pile will be outlined in the lawn and garden chapter of this book.)

What Can and Can't Be Recycled: A Quick Summary

What can and cannot be recycled depends in large part on what your local recycling industries will and won't accept. There are many items that can be recycled, even foam plastic cups and aerosol cans, but not every recycler has the technology or incentive to handle all items.

Another roadblock in what can and cannot be recycled is a market use for the recyclables. The paper industry could recycle more newspapers, but there aren't enough plants built to do the job. In addition, not all newspaper companies are using enough recycled newsprint to make it worthwhile for paper companies to spend the money to build the new plants.

New technology and new markets for recycled goods sparked by consumer demand need to come on line for recycling to reach full capacity and potential.

Here is a list of things that can and cannot be recycled.

PAPER

 Newsprint—can be recycled, but not the colored paper inserts.

Corrugated cardboard—can be recycled.

Computer paper and office stationery—can be recycled.

 Colored newspaper inserts, not even the funny papers.

Cereal boxes, food boxes, detergent boxes.

Magazines, telephone books, and junk mail.

Wax and wax-coated cellophane.

PLASTIC

 One and two-liter soda bottles, milk jugs, vinyl siding, antifreeze containers, motor oil and other auto fluids in hard plastic, hard plastic dish and laundry detergent bottles, ketchup and condiment bottles in hard plastic, shampoo bottles, baby wipes containers, Clorox and other bleach bottles, salad oil bottles, and most hard plastic containers.

 Cottage cheese and margarine tubs, plastic bags, and foam plastic containers (these are beginning to be recycled but most recyclers won't take them at this time).

GLASS

 Clear, brown, or green glass bottles and containers. No need to remove the labels.

 Windows, plates, light bulbs, casserole dishes.

METALS

 Aluminum and tin cans, clean aluminum foil, clean aluminum food trays, aluminum siding.

 No bottle caps or metals mixed with paper or plastic such as frozen fruit juice containers.

2

HOUSEHOLD
HAZARDOUS WASTE

Most people use household cleaners, paints, bug sprays, and flashlight batteries every day and have never once considered them to be dangerous. We've grown up using them. Our parents used them and nothing ever happened. Until now.

A surprising number and variety of common household products are classified as household hazardous waste. Some are toxic and poisonous, some corrosive, some flammable, and some even explosive. These products are not only hazardous to your health if swallowed or if they touch your skin, but they are hazardous to the air we breathe and the water we drink if they are not properly used, cared for, and disposed of.

Many of our most mundane items contain solvents, acids, alkalis, and a slew of chemicals from benzene to ammonia, and from cadmium to malathion. A chemical soup of potentially hazardous, toxic, and corrosive waste is cooking under your kitchen sink, in the cleaning closet, in the basement, and elsewhere in your home.

It is estimated that a typical American city of 100,000 persons dumps 3.75 tons of toilet bowl cleaner and 13.75 tons of liquid household cleaners down the drain each month. Households in the state of Washington alone put 14.7 million pounds of hazardous waste in their trash cans each year. These figures do not include the amount of hazardous waste we pitch in our backyards and down storm sewers. Most of that gets dumped in the water supply or in the garbage can.

The problem with hazardous waste is its use and its disposal. Up until now, many of us simply poured unused portions of, say, paint thinner down the drain, onto the ground, or into a storm sewer. What else do you do with this stuff? We thought, "My little bit of paint thinner won't hurt anything, right?"

Wrong. Here's what happened. If you poured it onto the ground, rain may have washed it on into a stream somewhere, or it may have filtered down through the earth into an underground aquifer or into deep wells. If you poured it down the sink, it went into the city water treatment system, which is not capable of neutralizing it. Then it went back into the city water supply and back into your tap when you went for a glass of water. Remember, there is no such thing as a throwaway, anymore. This stuff keeps coming back.

If you tossed the can with an unused portion of thinner in the trash, it probably went to the dump, which may not be very modern. The can opens when the bulldozer runs over it. The thinner spills out and drips down into the water supply and there it is again.

In this chapter, we are going to explain just what is household hazardous waste. We will identify specific items that are hazardous, tell you how to use and dispose of them, and give you alternative items to use. You will also learn how to find a permanent household hazardous waste station—there are over 1,800 in forty-four states. Or how to start one in your area if there is not.

(Motor oil, additives, lubricants, and auto batteries, plus cosmetics and lawn chemicals, are all household hazardous wastes, but they will be discussed at length in their respective chapters.)

WHAT ARE HOUSEHOLD HAZARDOUS WASTES?

Household hazardous wastes are those products that when discarded can pose a health and safety threat to humans, wildlife, and the environment. They can be liquids, solids, or gases. The Environmental Protection Agency says that hazardous wastes can be:

Toxic—which means if ingested, they can be poisonous, cause birth defects or cancer, or other health problems.

Corrosive—which means they can burn, eat away at and damage human flesh or even metals.

Flammable—which means they can catch fire or even explode.

Reactive—which means they can catch fire or explode if in contact with water.

It is not uncommon for household wastes to be toxic, corrosive, and flammable.

Now, some of the common categories of hazardous waste are solvents, asbestos, toxic metals, acids, alkalis, and pesticides.

Solvents. Solvents are the liquid substances in which other substances are dissolved. When you put sugar in water to make it dissolve, water is the solvent. Alcohol is a common solvent and so are turpentine and petroleum.

Many modern products are based on solvents, such as nail polish remover, paint thinner, dry cleaning fluid, paints, polishes, glues, and adhesives.

Many solvents are flammable, many are caustic, many

are toxic. All chemical solvents should be treated as hazardous waste and disposed of properly.

Acids. Acids in and of themselves are not necessarily bad. Lemon juice is acid and so is vinegar. Both are harmless. But strong, undiluted acids, such as battery acids, are dangerous.

Alkalis. Alkalis are just the opposite of acids, and the two should never be mixed. Never mix, never worry. Lye (sodium hydroxide), oven cleaners, drain cleaners, and ammonia are alkalis and they can be dangerous.

HOW TO MANAGE HOUSEHOLD HAZARDOUS WASTES

1. Read the label. Make sure you are getting the product you need to do the job you want.

2. Buy only the amount you need. Leftover products are the most likely to end up in the trash. You don't need it anymore, you don't want to be bothered, you toss it. It happens all the time in our busy lives.

3. If you have extra, ask your neighbors, friends, church, or civic organization if they need it and give the extra to them.

4. Use it up quickly. Once a product gets old, it may deteriorate or you may lose interest in using it. You may think it is no good and buy more, creating leftovers.

5. Use as directed. Don't use too little because it may not do the job. Don't use too much because it could harm you or the environment.

6. Use alternatives. Alternatives will be listed later in this chapter.

7. Recycle it. Many products can be used twice. Take paint thinners. You can let it settle in the container, strain off the clear liquid and use it again.

8. Take it to a hazardous waste collection center.

Toxic metals. Mercury, found in some batteries used in watches, hearing aids, cameras, and smoke detectors, and cadmium, found in rechargeable flashlight batteries and in some paint pigments, are toxic metals.

HOW TO FIND A HAZARDOUS WASTE COLLECTION CENTER

Contact your local officials, mayor, sanitation department, county supervisor, state department of environmental protection or state department of conservation or natural resources. At least one of these offices will know if there is a permanent site in your state or area or if there is a temporary site that will be receiving waste in the near future.

HOW TO START A HAZARDOUS WASTE COLLECTION CENTER IN YOUR COMMUNITY

First realize there is a big difference between a temporary and a permanent hazardous waste collection center. These collection programs are very expensive to operate. You have to hire professional chemists or technicians who can identify your hazardous waste. Then you have to hire professional hazardous waste haulers to barrel it up and cart it away.

Chances are that only the larger cities will be able to afford a full-time permanent hazardous waste collection center. A temporary center that collects waste once or twice a year will probably be the best bet for smaller communities.

Here's what to do:

1. Contact your local city, county, and state officials and tell them that you want to help start a collection program.

2. Get in touch with civic, business, church, or

social groups and enlist their support. You will be much more effective if you are working with a group rather than on your own.

3. Make an appointment to talk with your state hazardous waste agency, usually the EPA or departments of natural resources or conservation. They will have information you can use to get your project under way.

4. Hazardous waste collection days are expensive. Raise money by:

▶ Asking state and federal agencies if there are any grants or matching funds.

▶ Asking local business or civic organizations if they have money to give.

▶ Figuring out a charge that can be collected from people when they dispose of their waste.

5. Add a waste exchange element to your collection day. For example, people who need paint can come and get it from people who have unused paint to dispose of.

RESOURCES

GSX Chemical Services, Inc., P.O. Box 210799, Columbia, SC 29221. (800) 845-1019. National hazardous waste handler will help you set up a collection day and provide you with information.

The Institute of Chemical Waste Management, 1730 Rhode Island Avenue, NW, Washington, D.C. 20036. (202) 659-4613.

HAZARDOUS WASTE "HOT SPOTS" IN YOUR HOME

There are six main hazardous waste "hot spots" in your home. Three of them, bathroom, car and garage, and garden, will be discussed in their own separate chapters. This chapter will discuss the:

1. Kitchen—mostly cleaners and polishes, but also food.
2. Workshop—paints, strippers, glues, etc.
3. Pesticides—sprays and powders, flea collars, etc.
4. Miscellaneous hot spots—batteries, pool and photo chemicals, etc.

Hot Spot Number 1—Under the Kitchen Sink

Aerosol cans. Ignitable and can contain residual chemicals. Use it up or take it outside and spray out the contents into the air. Place it in trash bound for landfill. Do not burn the can, as it might explode.
Alternatives. Use pump-action sprayers or simply apply item to cloth and wipe.

Air fresheners. Toxic and irritant. Contain methylene chloride, naphthalene, formaldehyde. Air fresheners do not freshen air, only mask it with oily film in your nasal passages. Solid air fresheners may be poisonous to children and pets. Use it up outdoors or give it to a friend. Place in trash bound for landfill.
Alternatives. Potpourri, scented candles, open windows and let fresh air in.

Ammonia-based cleaners. Irritant, toxic, and corrosive. Contains ammonia, ethanol. Use up these products and throw empty container in trash. Pour small amounts of unused portions in toilet and flush with lots of water.
Alternatives. Mix vinegar and salt for surfaces and baking soda and water for tub and tile. Try soap and water or lemon juice.

Bleach-based cleaners. Corrosive and toxic. Contains sodium or potassium hydroxide, hydrogen peroxide, sodium or calcium hypochlorite. Use it up or wash unused portions down drain with lots of water. Septic tank homes must use extra water to flush away bleach cleaners.

Alternatives. Use borax for household chores. Use dry bleach in the laundry.

Dish-washing and automatic dishwasher detergent. Neither is very hazardous. Use wisely and don't ingest.

Disinfectants. Corrosive and toxic. Contains diethylene or methylene glycol, sodium hypochlorite, pine oil. Use it up or flush it with lots of water and place in trash. If in aerosol, use it up or spray it outside in open air. Throw can in trash.

Alternatives. One-half cup borax and one gallon warm water.

Drain cleaners and drain openers. Corrosive and toxic. Contains sodium or potassium hydroxide, sodium hypochlorite, hydrochloric acid, lye, and sulfuric acid. Use it up as intended or take to hazardous waste collection site.

Alternatives. Plunger, metal "snake," avoid clogs by keeping hair and grease out of drains, flush weekly with boiling water, pour one-half cup baking soda and two ounces of vinegar in clogged drain, let settle, follow with boiling water.

Furniture and floor polish. Flammable and toxic. Contains petroleum distillates, nitrobenzene. Use it up carefully or save for hazardous waste collection day.

Alternatives. Use lemon juice and vegetable oil for polish. Mop your linoleum floor with mild detergent or vinegar and water. Mop your wood floor with a vegetable oil like Murphy's oil soap.

Oven cleaner. Corrosive and toxic. Contains ammonia, potassium or sodium hydroxide. Use up and dispose of container in trash.
   Alternatives. Baking soda and water.

Rug and upholstery cleaners. Irritant, corrosive, toxic. Contains naphthalene, perchloroethylene, oxalic acid, diethylene glycol. Save for hazardous waste collection day.
   Alternatives. Sweep and vacuum the rug and furniture. Sprinkle with dry cornstarch, baking soda, borax, or cornmeal and vacuum.

Scouring powders. Corrosive, irritant. Not very harmful, but does contain chlorine. Use it up or wrap tightly in a bag and dispose in trash.
   Alternatives. Use a nonchlorine scouring powder or baking soda.

Silver polish. Corrosive and toxic. Contains acidified thiourea, sulfuric acid, and petroleum distillates. Wrap tightly, place in another container and put in the trash.
   Alternatives. Soak in boiling water with baking soda and a piece of aluminum.

Hot Spot Number 2—The Workshop or Basement

Your biggest problem here is with the chemicals used in making oil-based paints, thinners, strippers, and preservatives. The EPA estimates that 27 percent to 43 percent of all household hazardous wastes disposed of in landfills were paint products.

Enamel or oil-based paints. Toxic and flammable. Contains pigments, petroleum distillates, and mineral spirits. Use it up or give it to someone who will. Take unused portions to hazardous waste collection.
   Alternatives. Switch to water-based latex paint.

Latex water-based paint. Generally not hazardous, but it still contains pigments. Use it up or give it to someone who will. Tightly closed cans of latex paint will last several years as long as they don't freeze. Unused portions can be left outside with the lid off to dry and harden, then wrap in paper or plastic and throw in trash.

Alternatives. No need to use alternatives.

Furniture strippers. Toxic and flammable. Contains benzene, acetone, methyl ethyl ketone, alcohols, xylene, toluene, methylene chloride. Use it up or give it to someone who will. Save for hazardous waste collection.

Alternatives. Sandpaper or use a heat gun.

Paint thinners and turpentine. Flammable and toxic. Contains n-butyl alcohol, acetone, methyl isobutyl ketone, petroleum distillates. You can recycle thinners and turpentine by rinsing your brushes in them, then let the fluid rest. The paint particles will fall to the bottom and you can pour off the clear thinners for reuse. The paint that settled out should be carefully stored and saved for hazardous waste collection.

Alternatives. Switch to water-based latex paints, which wash up with soap and water.

Wood preservatives. Flammable and toxic. Contains chlorinated phenols, mineral spirits, creosote, copper or zinc naphthenate. The government banned the use of wood preservatives containing creosote, arsenic compounds, and pentachlorophenol in 1986 because these items were too dangerous for home use. If you have any of these older wood preservatives, take them to a hazardous waste collection center. Never burn wood that has been treated because the toxic chemicals can be released in the fumes to be inhaled by you.

Alternatives. Buy wood that has been treated by the manufacturer. Seal it with shellac.

Hot Spot Number 3—Pesticides

Bug sprays, ant and roach killers. Toxic. Contains organophosphates, carbamates, and pyrethrins. Look at the label and do not use these banned products: aldrin, chlordane, DBCP, DDT, dieldrin, heptachlor, lindane, kepone, Mirex, Silvex, 2,4 5-T, toxaphene. You can use these products: dichlorvos, resmethrin, pyrethrin, carbaryl, diazinon, Baygon, malathion, metaldehyde, but use them strictly following manufacturer's suggestions on the can. Store unused portions and save for hazardous waste collection.

Alternatives. Keep your house and kitchen clean and all food put away in safe places. Use traps for roaches. Place chili powder at point of entry to repel ants.

Flea and tick sprays and collars. Toxic. Contains carbamates, pyrethrins, organophosphates. Use them up carefully, wrap tightly and place in the trash.

Alternatives. Try herbal collars and keep house well vacuumed.

Rat and mouse poisons. Toxic. Contains brodifacoum coumarin (e.g., warfarin), strychnine. Use them up carefully or save for hazardous waste collection.

Alternatives. Live traps. Keep house clean and food safely stored.

Hot Spot Number 4—Around the House

Artists' supplies. Toxic and flammable. Can contain pigments, glues, solvents, lead glazes. Use up your art supplies and keep out of the unsupervised reach of children. The Arts and Crafts Materials Institute has a nontoxic art supply certification program. Look for labels that say "CP" for "certified products," or "AP" for "approved products," meaning they are nontoxic.

Alternatives. Switch to water-based paints, inks, and

glues. Avoid toxic pigments, epoxy, instant glue, permanent felt tip markers, instant *papier-mâché*.

Batteries. Corrosive, toxic, ignitable. Dry cell and disk or button batteries found in flashlights, watches, cameras, hearing aids, and toys contain toxic metals like lead, mercury, cadmium, and others. If they crack or break, they can leak their contents into the atmosphere. Ask your jeweler, camera shop, or hearing aid store owner if they will recycle your spent batteries. Otherwise, save for hazardous waste collection or get a recycling program started in your area. Don't just throw them away.
   Alternatives. Use rechargeable batteries if possible.

Mothballs. Toxic. Contains naphthalene, paradichlorobenzene. Use them up or save them for hazardous waste collection.
   Alternatives. Cedar chips, sachets with dried tansy or lavender flowers.

Photographic chemicals. Corrosive, toxic, irritant. Contain acids, silver, hydroxide, nitrates. Use carefully or save for hazardous waste collection.
   Alternatives. None.

Pool chemicals. Toxic, corrosive. Contains sodium hypochlorite, algicide, muriatic acid. Use them up carefully or save for hazardous waste collection.
   Alternatives. Ultraviolet light systems.

3

THE
FAMILY AUTOMOBILE:
Dangerous Polluter

The *Exxon Valdez* oil spill off the coast of Alaska in 1989 should give us a good idea of how destructive petroleum products bound for our automobiles can really be. The good old family car can be a dangerous polluter. Almost every product that goes into a car, from gasoline, to motor oil, the air conditioner and battery, even the tires, is a potential pollution problem that has to be handled carefully.

But, luckily, some auto products, oil and batteries, can be recycled and new technologies are being developed to handle the ozone-depleting CFCs used in auto air conditioners. We will give you the specifics on those developments in this chapter. Also, we will tell you how to handle the hazardous waste your car produces, and give you tips on energy-efficient driving that cuts down on exhaust pollution, gets you better gas mileage, and extends the life of your car and its parts.

MOTOR OIL

Americans use 1.2 billion gallons of motor oil in vehicles each year. Half of that, or 600 million gallons, is burned up in normal engine use. The other 600 million gallons are drained when you get your oil changed. Half of you change your own oil, meaning that the environmental fate of 300 million gallons of used motor oil is in your hands.

The American Petroleum Institute estimates that as much as 240 million gallons of used motor oil are dumped, some estimates are even higher. This is a disaster for two reasons.

1. Recycled motor oil is a valuable resource. If the amount of oil we dump was burned in a power plant, it would create enough electricity to supply 360,000 houses for a year. It takes forty-two gallons of crude oil to refine two and a half quarts of motor oil. It takes only one gallon of recycled oil, and half the energy costs, to create the same two and a half quarts of motor oil.

And motor oil can be recycled. It is law in most states that any service station or quick oil change shop must receive used motor oil from any source. Many stations are reluctant to take used oil because they are being forced to pay companies to haul it away, but they are nonetheless required to do so.

Fifty-seven percent of recycled motor oil is used as fuel oil, 26 percent is used to make new motor oil and other lubricants, 17 percent is used as road oil, dust control, wood preservatives, and "fire log" ingredients.

2. Dumped motor oil kills wildlife, pollutes lakes, streams, and oceans, and will eventually end up in your drinking water. Just one quart of motor oil can contaminate up to 2 million gallons of drinking water.

The four quarts of oil you just drained from your crankcase can form an oil slick nearly eight acres across.

WHAT CAN I DO TO RECYCLE MOTOR OIL?

1. Collect your waste oil and take it to a certified waste oil collection site.

2. If you are having problems with service stations that are reluctant to take your used oil, start an oil-recycling center with your city or town highway department.

3. Place your used oil in a closed container and label it carefully.

4. Don't mix with other substances like gasoline, paint strippers, or pesticides in your used oil.

5. Call your state or federal EPA for help: (800) 424-9346.

CAR BATTERIES

Wet cell batteries, those found in your car, truck, boat, or tractor, contain lead, which causes cancer, and sulfuric acid, which is highly caustic. If you just toss a used battery in a landfill, eventually it will crack and the lead and sulfuric acid will leak into the ground and water supply. Even if you leave it in your garage, tool shed, or in the basement, eventually it will crack and cause damage.

You can recycle your batteries. Take them back to the place where you bought your new replacement battery or take them to a recycling center. If you can't recycle your battery in your area, save it for a hazardous waste collection day. Pack it in heavy-duty hard plastic and keep it covered until you can get rid of it.

AUTO AIR CONDITIONERS

Air conditioners, even the one in your automobile, use chemicals called chlorofluorocarbons, or CFC-12. The release of CFCs into the air and the earth's atmosphere has been linked to causing the worldwide warming trend and the greenhouse effect, and to stripping away the earth's ozone layer.

Forty-one percent of the CFCs used for air conditioning are used in automobiles. But auto air conditioners are very leaky machines. So, 75 percent of replacement CFCs are used by auto air conditioners. The EPA says that automobile air conditioners are the single largest use of CFCs in the United States. The United States is the only country in the world that installs air conditioners in automobiles on a regular basis.

A big problem, besides leaking, is that auto air conditioners need to be serviced quite often. Typically, the old air-conditioning gas is allowed to leak out by the service station operator, releasing dangerous CFCs into the environment.

But, recently, new technology has become available to allow service station operators to capture and recycle the used refrigerants rather than venting them into the air. The

---

WHAT CAN I DO?

1. Try using your auto air conditioner less. Turn the temperature gauge down just a little. Open the windows and let in the fresh air.

2. Encourage your service station operator to buy and use one of the new systems that allow them to recycle used refrigerant. They cost between $2,500 and $7,000.

3. Go to a service station that has one on line.

4. Air conditioners in old cars in junkyards are a potential problem. Salvagers could be encouraged to suck out the old refrigerants and sell them to service centers to use as recycled materials.

---

state of Hawaii has mandated that service centers use these recovery systems when servicing auto air conditioners.

TIRES

Used auto tires present a peculiar garbage problem for the United States. Only 1 percent of our garbage is tires, but that translates as 2,500,000 tons, or, better yet, one tire per person per year.

Tires take up a lot of room in landfills. They are round, with a big hole in the middle, and they have to be stacked. Another feature of tires is even if you bury them, they work their way to the surface.

Because of a shortage of landfill space, many states have banned the dumping of tires. These bans have caused illegal tire dumps or piles of tires to spring up. State officials in Ohio alone estimate there are more than 28 million scrap tires stockpiled there.

Tire piles catch fire and attract mosquitoes. The average scrap tire contains two and a half gallons of petroleum, and when it burns it releases this as oil, which seeps into groundwater. Tire piles give off huge billows of toxic smoke when they burn, and they are very difficult to control and extinguish.

Scrap tires outdoors collect water and other organic waste materials, ideal conditions for mosquitoes. Mosquitoes carrying yellow fever love to breed in tires.

WHAT CAN BE DONE?

1. Extend your tire wear. Keep tires properly inflated and rotate your tires according to the manufacturer's specifications. Make sure your tires are balanced and your wheels are aligned.

2. Buy retreads. Approximately 50 percent of scrapped tires could be retreaded.

---

3. Incineration. Burning tires in power plants creates more energy than coal.

4. Recycling. Tires can be stamped into rubber mats, used as boat dock bumpers, shoal and reef along ocean shores, highway crash barriers, and used in combating soil erosion.

5. Shredded tires. Shredded tires can be remade into roofing materials, athletic surfaces, asphalt rubber, and other items.

AUTOMOBILE FLUIDS

Antifreeze. Toxic. Contains ethylene glycol. Three ounces of antifreeze ingested can kill an adult. Antifreeze spilled during servicing often kills pets who drink it because of its sweet flavor.

People typically flush the antifreeze out of their radiators with water. Do not let this antifreeze and water solution go into a storm sewer, on the ground, or into your septic tank. Collect it and flush it down the drain hooked to a sewage treatment facility.

Automatic transmission fluid. Toxic. Contains mineral oil. Used fluid contains lead and other heavy metals.

Place drained transmission fluid in a plastic container with a tight-fitting lid. Bring for recycling to a service station or highway department that receives waste oil.

Brake fluid. Toxic and flammable. Contains glycol ethers and heavy metals such as lead. Cannot be recycled or disposed of at home. Must be taken to a hazardous waste collection center.

Car wax. Corrosive and flammable. Contains caustics, acids, petroleum distillates. Use up the car wax you have or give it to a friend. If it is hard, you can wrap it in plastic and throw in trash.

Gasoline. Highly toxic and flammable. Contains petroleum distillates, lead in leaded gas, and high-octane components benzene, ethylene dichloride, and methanol.

Usually gasoline is stored in tanks and burned in your engine. Unused gasoline in cans is a problem. Do not dump it on the ground. Save for hazardous waste collection day.

Windshield washer solution. Flammable. Contains methyl alcohol, methanol. Save unused portions for hazardous waste collection day. The methyl alcohol in washer fluid is there to prevent the fluid from freezing in cold weather. In warmer weather, use plain water.

ENERGY EFFICIENT DRIVING—OR, HOW TO CUT DOWN ON THE AMOUNT OF EXHAUST MY CAR DUMPS INTO THE AIR AND GET MORE MILES PER GALLON WHILE DOING IT

Since unleaded gas and cars needing unleaded gas plus the introduction of the catalytic converter, pollution from cars has decreased. Until recently. Now there are so many more cars on the road that pollution is again on the rise. From 1977 to 1988, the number of cars rose 25 percent. There are now over 183 million cars on the road, plus 41 million trucks.

Cities and states are beginning to take action to limit the number of cars. California has mandated that businesses find new ways of reducing the use of cars, or face heavy fines.

Car manufacturers continue to improve the energy efficiency of engines. Oil companies are reformulating their fuels to burn cleaner. New technologies for using natural gas and methanol in cars and buses are being tested.

But, in the meantime, there are many things you can do as a motorist to use gas wisely and cut down on auto air pollution. You can become an energy efficient driver and get 30 to 50 percent better mileage from your car. Here's a list.

1. Use unleaded fuel in cars that call for it. As much as 20 percent of leaded fuel from pumps is being used in unleaded cars. This fouls your catalytic converter and spews lead into the air.

2. Avoid excessive warming. Let your car warm up for no more than thirty seconds. After that, you are wasting gas and dirtying your engine.

3. Don't sit and idle your car when not in traffic. Turn the engine off and restart it when you are ready to go again.

4. No hot rod driving. Jackrabbit starts, drag racing, and squealing tires waste gasoline.

5. Obey the fifty-five-mile-per-hour speed limit and you'll get 21 percent better mileage than at seventy.

6. Forget what the Shell Answer Man said about placing an egg under your accelerator. You use more gas during acceleration because you are overcoming gravity. Remember? Bodies at rest tend to stay at rest. Bodies in motion tend to stay in motion. The laws of gravity. Get your car up to speed quickly and maintain a steady speed.

7. Avoid short trips. It's better to run a lot of errands all at once because your car performs better once it is warmed up. Don't make a short trip, return home, and then make another short trip.

8. Keep your car maintained. Keep tires inflated to proper pressure. Change spark plugs. Have your car serviced on a regular basis.

THE LAWN
AND GARDEN

Just take a look outside at your beautiful lawn and garden. Isn't it lovely? All nice and green and well manicured. No pollution here, right?

Wrong. Dead wrong. Your yard, mainly in the form of grass clippings and fallen leaves, is the second largest pile of garbage that goes to the city dump. As much as 20 percent of landfill space is taken up by yard waste.

Furthermore, in caring for your lawn, you dump tons of chemical fertilizer, fungicides, pesticides, and herbicides on your lawn, much of it extremely dangerous toxic chemicals, a lot of it contributing to polluted waterways and fish that we are warned not to eat.

In this chapter, we are going to show you how you can stop the destruction of clean water and healthy wildlife by growing a chemical-free lawn. We will tell you how to compost those leaves and grass clippings with food waste from the kitchen and not smell up the neighborhood.

FALLEN LEAVES AND GRASS CLIPPINGS

Fallen leaves and grass clippings make up about 15 percent to 18 percent of yard wastes. The rest comes from brush and limbs, but we will talk about them later. During the summer mowing season many of us bagged our grass clippings and placed them on the curb for pickup. Years ago, you could simply rake your leaves and burn them in a smoky autumn ritual. You can't do that now in most areas because the smoke creates too much air pollution.

During the fall raking season, leaves can account for 35 to 50 percent of the material dumped at the municipal landfill. We know that landfill space is at a premium and will soon run out in many areas. Besides, all these leaves and grass clippings piled together and buried start to decompose, creating the buildup of explosive methane gas and contributing to groundwater leachate. Many states and cities are now banning the dumping of fallen leaves and grass clippings.

So what are we supposed to do with this stuff? The good news about grass clippings is that you can simply leave them on the ground. They decompose rapidly, feed the lawn naturally, and do not cause thatching. If you really feel the need to gather your clippings and remove them from the lawn, you can do the same thing to them as you can do to leaves, COMPOST THEM.

Throughout the heavily populated Northeast, municipal composting of leaves and grass clippings has recently become a way of life. Some municipalities send special trucks around to collect the bagged leaves or clippings. Some require homeowners to take the debris to collection stations.

Leaves and grass clippings are being increasingly composted, which means they are placed in huge piles and left to decompose naturally in the open air, which turns them into compost, a nutrient-rich brown, crumbly substance that is very much like garden soil. It is then hauled away and used as fertilizer or soil builder on farmers' fields and residential lawns.

But don't these leaves and grass clippings have to be placed in bags? Yes. And the answer to that are the new biodegradable bags that are coming on the market. The plastic in these bags is infused with cornstarch or other substances that break down in the compost environment and cause the bags to break down, too. This allows the leaves and grass clippings inside to break down and decompose, too.

The Archer Daniels Midland Company (ADM) in Decatur, Illinois, is the maker of the cornstarch that causes the compost bags to decompose. You can also buy huge biodegradable paper bags at some lawn and garden centers or by mail order from Seventh Generation, 10 Farrell Street, South Burlington, VT 05403. (802) 862-2999. Catalogue, $2.

But you don't have to rely on municipal collection services to compost your leaves or grass clippings. You can build your own compost pile in your own backyard that does not smell and will provide you with plant food and nutrient-rich natural fertilizer for your lawn and garden.

HOW TO BUILD A COMPOST PILE

Your compost pile will be approximately three feet by three feet and three feet high. It can be a square or a circle. So look for a place in your yard that is not conspicuous, that won't be in the way of your family fun and entertaining, that is easy to get to, that you feel comfortable using for a compost pile. It doesn't make much difference if your pile is in the sun or shade, as long as there is plenty of air space all around it. If you have a choice, place it in the sun. Don't site it up against your house, garage, or shed. Your site needs to be level and well drained.

Build your compost bin out of whatever materials are

handy or cheap. Chicken wire, storm fence, fence wire, pallets, used lumber—you can even use concrete blocks or bales of hay if you have them around. Just make it three feet by three feet by three feet and make sure there are openings on the sides large enough to let air in but not so large that the compost spills out.

Filling the compost bin is easy. Just fill it up with leaves, grass clippings, leftover tomato vines, dead flowers, whatever. But the best compost piles, the ones that really work, have a combination of items high in carbon, like leaves, shredded tree limbs, twigs, and paper, and items high in nitrogen, like grass clippings, table scraps, corn wrappings, fruit peeling. The best ratio for a fast-working compost pile is twenty-five times as much carbon as nitrogen. But it is better to build a compost pile with what you have rather than worry about carbon-to-nitrogen ratios.

A NOTE ABOUT KITCHEN FOOD SCRAPS

Food wastes account for 8.4 percent of garbage dumped in our landfills. If you can take your potato peelings, apple cores, coffee grounds, leftover salad, heels of bread, stale potato chips, almost everything you eat or drink, to your compost pile, you will be doing the world a great environmental favor. It is possible to place bones and meat in the compost pile, but they decompose much more slowly and they are the culprits in attracting the neighborhood dogs, cats, and rodents to the pile.

Water your compost pile thoroughly when you first get it started. Let rain do the rest unless you go through an unusually dry period.

Take a pitchfork or forked spade to the compost pile and turn it once a week or once every two weeks. The compost is literally "cooking" in there and you want the cooler outside stuff to be blended in with the hotter inside stuff. It is much the same as stirring a vat of stew.

You don't have to turn a pile if you don't want to. Turning moves the process along faster. You should get ready to use compost in two to six months, depending on how often you turn it and how much nitrogen you put in the pile. You can buy compost additives, such as Compost Plus, that help activate the process and speed it along.

You can also buy specially designed prefabricated composting bins that allow you to put your organic substances in the top and draw the compost out the bottom—no fuss, no muss.

Gardener's Supply (128 Intervale Road, Burlington, VT 05401. [802] 863-1700) has the most comprehensive collections of garden and lawn composting tools, supplies, and equipment for sale in a catalogue. Everything from compost fences, to special shovels, to activators, worms, even the Cadillac of composters for $99.95. They will send you a free copy of *Simple Steps to Successful Composting.*

BRUSH AND TREE LIMBS

Brush from bushes and clippings from bushes and tree limbs pose certain problems for composting, either municipal or residential. Some landfill areas are simply open-burning these items rather than placing them into the landfill itself.

But that is a mistake. If shredded finely enough, these items can be composted successfully. But that's not necessary. Shredded brush and smaller tree limbs make excellent mulch to be applied to gardens and flower beds. Mulch acts as a vapor blanket, which lets water into the soil beneath and holds it there during the ravages of hot summer days. It also is an excellent weed preventer, much better than herbicides.

Shredders are expensive, priced from $400 to $1,500 for residential uses. They can also be dangerous if you don't know how to use them correctly. But a block or neighborhood association could afford to buy one and share the use of it among all. But, better yet, towns and municipalities could buy industrial-strength shredders, like the "tree eaters" used by power and telephone companies to keep their lines clear.

Furthermore, larger limbs could be cut with a chain saw and used for firewood or incorporated into the ground for flower garden borders. But chain saws are again dangerous tools in the hands of a novice. A better solution is for the town to hire a chain saw crew to come around to homes and cut the wood up on location. They could haul the firewood away or leave it for your use. Wood hauled away could be sold to pay for the wages of the crew. Private entrepreneurs could also start businesses along these lines and create jobs and profits.

THE CHEMICAL-FREE LAWN

It is estimated that there are 20 million acres of lawn in the United States. In recent years, the amount of money people have been spending on their lawns has been increasing. That means we are not only buying more equipment and plants from greenhouses, but applying more fertilizer, pesticide, and herbicide. We dump from five to ten pounds per acre of pesticides on our lawns each year.

The key to a chemical-free lawn is to make your lawn a healthy lawn. You need good soil, the right kind of grass seed, proper mowing and watering. If you depend on chemicals to grow grass and kill weeds and pests, your lawn will become dependent on those chemicals and not be healthy. Sounds familiar, doesn't it? Just say no to chemicals, even on your lawn. If you use chemicals, use them wisely, and follow instructions carefully.

FERTILIZER

Most people buy chemical fertilizer and spread it on their lawn with little machines called "fertilizer spreaders." Fertilizer is made up of nitrogen, phosphorus, and potassium. A number like 5-10-5 corresponds to the percentage of each nutrient, by weight, in that bag.

People put fertilizer on their lawn to make the grass grow greener, taller, and thicker. Unfortunately, too many people think that if they apply twice as much fertilizer as the package recommends, the results will be twice as good.

Wrong. It is estimated that as much as half of the soluble nitrogen we apply to our lawns will be washed off and never do its job. Which brings up a very important pont about fertilizer. When fertilizer is improperly applied and runoff occurs, the nutrients go into the water supply and cause problems.

Too much nitrogen in water causes oxygen starvation, which kills fish, and can cause blue baby syndrome in young children if they drink it.

Too much phosphorous in water results in algae blooms, the famous red and brown tides on the East Coast that are killing the clams, scallops, and oysters that live there.

You can't just dump fertilizer on your lawn. Too much or too little at the wrong time can actually damage your lawn. Here's how to do it right:

1. Have your soil tested. You can buy inexpensive kits to do this yourself or contact your country cooperative extension agent for help. Extension agents are everywhere, even New York City, which hasn't had a farm in some time.

2. Plant the right kind of grass seed. Certain grasses do better in certain parts of the country under shady or sunny, wet or dry, conditions.

3. Clean up your yard. Rake the leaves, twigs, cigarette butts, candy wrappers, and other debris from the lawn.

4. Mow your lawn to the proper height and at certain intervals. Different grasses need different heights. Never mow off more than one-third of its height. If you want grass to be two inches tall, mow it before it reaches three inches. Leave your grass an extra inch tall during dry spells and in shady areas.

5. Use nonchemical fertilizers. Sources:

Nitrogen. Fresh or dried cow or chicken manure, blood meal, cottonseed meal, even some of that compost you just made, if it is finely sifted, will benefit your lawn.

Phosphorus. Bone meal or soft rock phosphate.

Potassium. Greensand, granite dust, tobacco waste, dried sheep manure, millet and buckwheat straw.

HERBICIDES

Weeds are plants growing in the wrong place. Herbicides are used to kill weeds. The first questions you should ask yourself are: "Are those few weeds really so bad that I have to dump poisonous chemicals on the lawn to get rid of them? Can't I just dig them out by hand or just ignore them?"

In the lawn, the best way to get rid of weeds is to create a healthy stand of grass. A healthy lawn will keep weeds out and not let their seeds germinate and sprout. If you do have a particular problem with a patch of dandelions, dig them up, loosen the soil in the patch, cover with some fresh soil or compost, and sprinkle with new grass seed.

In the garden, you can pull weeds up by hand or chop them with a hoe. Now, I admit, this can become tedious. Your best alternative is heavy applications of mulch.

Mulch is a layer of organic matter that acts as a barrier to smother weeds. It also keeps your garden moist and cool during heat waves and droughts. Simply apply the mulch three to six inches thick in between rows and in between plants. Don't apply mulch around tomatoes, peppers, and eggplant until the soil is warm.

Good mulches are fresh or rotted hay, grass clippings, shredded leaves, shredded tree bark and limbs, compost, rotted manure, sawdust, seaweed, even newspapers and black plastic.

PESTICIDES

There are two main types of pesticides. There is the chemical type and the so-called natural type. The natural type should not be considered safe by any means. All pesticides are toxic. They are poisons that if ingested in large-enough doses or over a long period of time can kill you and wildlife or make you both very sick.

All pesticides should be treated like household hazardous waste. Unused portions should be saved for hazardous waste collection. Never dump pesticides down the drain, dump them on the ground, or toss unused portions in the trash can.

The way to tell how dangerous a pesticide is is to look at the label. The EPA requires that one of these three words be on the label.

▶ DANGER or POISON—the most lethal. Means that a few drops to one teaspoon of this pesticide is dangerous.

▶ WARNING—second most lethal. Means that one teaspoon to one ounce of this pesticide is dangerous.

▶ CAUTION—third most lethal. Means that more than an ounce of this pesticide can harm your health.

Let's take a look at the different pesticides, learn how dangerous they are and how they should be used.

Chemical Pesticides

Many chemical pesticides are better known by their brand names. But, for you, it is better to know which chemicals are dangerous, and watch for them on the labels, than to memorize the different brand names.

Chlorinated hydrocarbons. DDT is the most widely known chlorinated hydrocarbon pesticide. DDT is banned now, so are aldrin, dieldrin, and chlordane. These pesticides attack the central nervous system to kill pests. One organochlorine that is still on the market and available to consumers is methoxychlor, sold under the brand name Marlate.

Organophosphates. Malathion, parathion, diazinon, dichlorvos, chlorpyrifos are organophosphates. Some of their brand names are Dursban, Lorsban, Cythion, Spectracide, and KnoxOut. These chemical pesticides also disrupt the pest's nervous system to kill. These are widely in home and garden use.

Malathion is a common ingredient in houseplant pesticide.

Carbamates. Carbaryl, aldicarb, carbofuran, and propoxur are carbamates. Brand names are Sevin, Temik, Furadan, and Baygon. These also act on the pest's nervous system to kill.

Natural Pesticides

Natural pesticides are derived from plants, that's why they are also called "botanical pesticides" or "botanicals." But just because they come from plants does not mean they are not toxic. Two of these, rotenone and nicotine, are many times more toxic than the organophosphate malathion or the carbamate Sevin. Don't let the word "natural" fool you.

Natural pesticides are considered better in the long run because they are usually mixed in very small doses when they get to the consumer, meaning they are less toxic than chemicals at the point of purchase. Next, they persist in the environment for shorter periods of time. Most chemical pesticides stay potent in the environment for a long time, many months, or even years in the case of DDT. Naturals tend to do their bug killing and then break down and disappear.

Nicotine. Most toxic of the botanicals. Made from tobacco plants. Most often sold as nicotine sulfate and known under the brand name Black Leaf 40. Banned in several states because of its toxicity.

Rotenone. Second most toxic botanical. Extracted from roots of a number of different plants. Sold under its own name as well as Prenox and Noxfish. Very toxic to fish. Do not use near waterways.

Ryania. Third most toxic botanical. Made from the ground stems of the South American shrub *Ryania speciosa.* Sometimes sold in a mix called Triple Plus (Natur-Gro). Persists longer than other botanicals.

Pyrethrum. Fourth most toxic botanical. Extracted from the seeds of a type of chrysanthemum. Relatively nontoxic to humans but should not be used near waterways, where it is more toxic to fish.

Sabadilla. Least toxic botanical. Extracted from seeds of South American lily. Sold under the name Red Devil Dust. Toxic to honey bees. Apply in the evening when bees are less active.

There are other ways to control pests in your garden

that do not include the use of either chemical or botanical pesticides.

Cultural controls. Means you create a garden or lawn that is so healthy, full of rich nutrients, well watered and cared for, clean and debris free, that insects cannot find a home there.

Mechanical controls. Pick the bugs off by hand. Place netting over lettuce or cabbage plants. Spray with insecticidal soaps that are totally nontoxic but cause bugs to dehydrate and die.

Biological controls. These are bacteria and natural predators that kill insects. *Bacillus thuringiensis* and *B. popillae*, better known as milky spore disease, are two examples. Ladybugs, praying mantises, snakes, and toads are other examples.

Here is a list of companies offering natural pest controls.

Green Earth Organics, 9422 144th Street, Puyallup, WA 98373. (206) 845-2321. Free catalogue.

Harmony Farm Service and Supply, P.O. Box 451, Graton, CA 95444. (707) 823-9125. Catalogue, $2.

Integrated Fertility Management, 333-B Ohme Gardens Road, Wenatchee, WA 98801. (800) 332-3179. Free catalogue.

Mellinger's Inc., 2310 West South Range Road, North Lima, OH 44452. (216) 549-9861. Free catalogue.

The Natural Gardening Company, 217 San Anselmo Avenue, San Anselmo, CA 94960. (415) 456-5060. Catalogue, $1.

Natural Gardening Research Center, P.O. Box 14, Sunman, IN 47041. (812) 623-3800. Free catalogue.

Nature's Control, P.O. Box 35, Medford, OR 97501. (503) 899-8318. Free brochure.

The Necessary Trading Company, New Castle, VA 24127. (703) 864-5103. Catalogue, $2.

Ringer, 9959 Valley View Road, Eden Prairie, MN 55344. (612) 941-4180. Free catalogue.

Safer Inc., 189 Wells Avenue, Newton, MA 02159. (800) 423-7544. Free catalogue.

5

SAVVY SUPERMARKET SHOPPER

Savvy shopping at the supermarket might just be one of the most productive things you can do to cut down on the amount of garbage you produce. Each person in the United States creates three and a half pounds of garbage per day. For a family of four, that's fourteen pounds of garbage every day.

Packaging accounts for 30 percent of that trash by weight, and almost 34 percent of that trash by volume. Food packaging by far is the biggest part of total packaging. It is estimated that you alone toss 440 pounds of packaging per year into your town's landfill. Next time you do a week's worth of shopping and drag home a dozen bags of groceries, take a look at the packaging. Separate and save those packages and you'll see just how much you produce, and how much you can cut down.

Paper makes up 48 percent of the packaging that gets thrown in the trash and heads for your already over-crowded and soon-to-be-full dump site. Glass is 25 percent of packaging; plastic, 13 percent; tin cans, 6 percent; aluminum cans, 2 percent; and about 5 percent miscellaneous.

In this chapter, we are going to take you on a tour of the average American supermarket. We are going to go up and down the aisles, from produce, to deli take-out, from the meat case to the frozen food section. We are going to show you how to make "green" decisions, or how to choose the best packaging that will have the least amount of environmental impact. We will point out several "packaging pollution problems" and discuss the alternatives, which are often already right there on the shelves or soon to be.

Let's start our journey at the end, at the checkout counter, where you have to chose between paper and plastic bags.

PAPER GROCERY SACKS OR PLASTIC BAGS— WHICH IS BETTER?

On the surface, most people would assume that the paper grocery sack is the better choice because, since it is made of paper, it is biodegradable and harmless. But the process used to create paper at mills can be very polluting to nearby water sources. Trees have to be cut down, even to make recycled paper.

Paper is also five times as heavy and ten times as bulky as plastic sacks, meaning that paper sacks take up more room at landfills.

On the other hand, plastic bags have their drawbacks. When people litter plastic sacks, and those sacks get into waterways, especially oceans, fish, birds, and turtles eat the plastic sacks thinking they are food. Thousands of animals die each year of starvation or suffocation linked to the eating of plastic bags.

If you bring your groceries home in either plastic or paper bags, empty the groceries, and then throw away the bags, you are simply polluting. Choose to recycle those bags. Many communities that recycle newspapers insist that the papers be brought either tied in bundles or packed in brown paper grocery sacks. A perfect use for them. Paper bags can also be used as scratch paper for children who love to draw or paint. Plastic bags can be used to wrap sandwiches. They are a little big, but at least you get two uses out of a bag before you discard it.

Paper bags and plastic bags are about equal in their environmental impact. So the choice is really up to you. You could carry a reusable heavy canvas bag for your groceries.

Before we get into all the specifics of supermarket shopping, let's lay down a few general rules that will help.

Rule Number 1. Buy in bulk whenever possible. You get more product and less package.

Rule Number 2. Buy containers that can be recycled. Glass, tin cans, aluminum, and some plastics can be recycled. Cardboard, unless it is corrugated, cannot.

Rule Number 3. Buy packages that are made of recycled paper. Most cereal boxes and many others are made of recycled paper. It usually says so on the box.

Rule Number 4. Avoid excess packaging. Many convenience foods are all dressed up in fancy packaging that is unnecessary and wasteful.

THE PRODUCE AISLE

Produce in its natural state is very nonpolluting. Problems occur when we put all of our different produce picks in plastic packaging. First there are the hanging rolls of plastic bags that are hard to open. Do you really need them? No, not really.

Buy only enough produce for the next two or three days, put it in the crisper, and it will be fine. Fruits, especially, do not need to be individually bagged, which is only a convenience to the supermarket so that they can weight and price each item. Just put your produce in the cart and take it to the checkout. Don't worry, they will sell it to you even if it is not individually bagged.

Sometimes, the grocer prepackages the fruits and vegetables for you. If loose oranges and bagged oranges are both priced the same, buy the loose oranges. It doesn't matter if the package is plastic wrap over paper or plastic wrap over plastic trays. It is excess packaging that you simply don't need.

DELI AND PREPARED-FOODS SECTION

This is the fastest-growing sector of the supermarket industry. You can buy whole meals here now. But, usually, they are heavily packaged, most of the time in plastic.

First, this plastic packaging is bulky and takes up a lot of room in the trash can. Second, in most cases, the plastic is not being recycled. One exception is the hard plastic containers that resemble Tupperware or Rubbermaid containers. Pasta sauces and soups are often packaged in these. These are high-density polyethylene plastic, or HDPE. These can be recycled and reused in the home.

Sliced meats and cheeses from the deli case do not need to be overly wrapped. Tell the person to wrap it in paper or plastic, but don't bother to double wrap it.

THE MEAT COUNTER

Almost exclusively now, meat and poultry are precut and packaged at supermarkets. It is only at the smaller markets and butcher shops that the meat is cut to order and wrapped for you.

Meat and poultry do not need to be nestled on a plastic tray and wrapped in plastic film. These trays create an enormous amount of trash. You will have a very difficult time getting the supermarket butcher to change his packaging, but you can try. Tell him you do not want the plastic tray. Encourage him to change his packaging to butcher's paper, which creates less trash. Or go to a different butcher who will.

TIN CANS, GLASS JARS, PLASTIC BOTTLES

Plastics are increasingly taking over in this area. It is hard to get even ketchup bottles that aren't made of plastic. The problem with some plastic containers is that they are made of commingled plastic. Which means that they can be difficult to recycle because they are made of different types of plastic.

The plastics industry is pedaling as fast as it can to change this and make plastic more recyclable. It is even trying to get containers to display symbols, which indicate if the container is recyclable and what type of plastic it is.

Milk containers are a good example of the plastic container dilemma. Paper milk cartons would on the surface appear to be more environmentally benign. But no. Those containers cannot be recycled because they are covered with wax. Plastic milk containers, on the other hand, are made of HDPE, which is recyclable.

In many towns, people are charged heavily for the amount of trash they produce that cannot be recycled. Paying extra to dispose of one type of empty container rather than recycle another is a good guide to use when making a purchasing decision.

Because glass can be recycled over and over again as food containers, it is right now the best type of container to buy.

Aluminum and tin cans are also highly recyclable. They are second best.

Plastic is becoming more recyclable every day. But most of the plastic that is recycled is not used in products bound for food packaging in the supermarket. Because of public health concerns, new plastic is still used for food packaging. There are some exceptions, so let's talk about them now.

LAUNDRY DETERGENTS

There are two problems with laundry detergents. One is the packaging, either plastic, paperboard, or cardboard. The second is their contents. If they contain phosphates or bleaches, they can contribute to water pollution.

Let's start with packaging. Recently, Procter & Gamble, Lever Brothers, and Colgate-Palmolive have started using varying amounts of recycled plastic in their plastic detergent bottles. Furthermore, most of those bottles are made with HDPE, a type of plastic that can easily be recycled.

Your general alternative is boxes of detergent. Many of those boxes have for years been made with recycled cardboard. Look for the recycled symbol on the package. Unfortunately, those boxes cannot be recycled, so they end up in the trash.

Corrugated cardboard, on the other hand, can be recycled. But you usually have to buy the huge twenty-five-pound economy-size box to get corrugated cardboard.

BOTTOM LINE. Buy detergents in the largest size you can afford. It usually is a better buy and you create less packaging waste.

BOTTOM LINE. Buy detergents in hard plastic containers because they are made of recycled plastic and are recyclable.

Many detergents contain phosphates, which act as a water softener to allow the cleaning agents in the detergent to do a better job. The problem is that phosphates are also a very rich plant food. Phosphates in detergents were banned on Long Island, New York, because the phosphates were washing into the water, becoming a food source for algae, and causing huge algae blooms, known as red and brown tides. These blooms killed millions of fish and shellfish and fouled the water.

Phosphates are not in detergents in states where phosphates are banned. But there are no phosphates in liquid detergents, only in the powdered ones.

BOTTOM LINE. Buy liquid detergents.

BEST DETERGENT BUY. Liquid detergents in giant-size plastic containers.

MICROWAVABLE FOOD AND SINGLE SERVINGS

Microwavable foods and foods packaged in small single-serving size are flooding your supermarket. They're convenient and designed to fit today's busy lifestyles.

They are also huge polluters because they are mostly excess packaging and very little food. These two categories should be avoided altogether if you want to protect the environment.

EXAMPLES OF POLLUTERS

Juice packs. Cannot be recycled because they are made of layers of paper, plastic, and foil. Buy a big can of juice and pour it in a glass. Recycle the can. Wash the glass.

Pudding snacks. Cannot be recycled. Make pudding and dish it out instead.

Frozen dinners. Neither the paper cover nor the plastic tray can be recycled. Old-fashioned aluminum trays can be recycled.

Campbell's Souper Combo. Soup and sandwich double- and triple- wrapped in paper and plastic. None can be recycled.

BOTTOM LINE. Avoid excess packaging.

PACKAGING POLLUTION PROBLEMS

Egg cartons. Foam plastic, known as Styrofoam, no longer is made with ozone-depleting CFCs. The plastics industry is beginning a few pilot projects to recycle foam plastic, but, in general, it is not recyclable right now. Paper cartons can be reused or torn into tiny pieces and added to your compost pile.

Coffee filters. Paper is made white by the use of chlorine bleach, which is a source of water pollution. Look for brown coffee filters, which are made of unbleached paper. Melitta makes them.

Toilet paper. Toilet paper made from recycled paper, not recycled toilet paper, is available and on the market. It can also be mail-ordered from Seventh Generation, 10 Farrell Street, South Burlington, VT 05403.

Household cleaning supplies. These products are discussed at length in the household hazardous waste chapter in this book. Most of these items contain dangerous chemicals, many of them toxic. Please consult that chapter for safe alternatives.

6 ▶

DIAPERS, THE LAUNDRY, COSMETICS, AND THE BATHROOM

We've clustered a group of seemingly unrelated environmental concerns under one umbrella in this chapter. But when you think about it, there is some connection. They all have to do with cleanliness, for ourselves, our families, and our environment.

Diapers are a major concern for many people. Are disposable diapers the right choice, or should we consider returning to cloth diapers? Are the detergents, bleaches, stain removers, and fabric softeners we use every day in the washing machine safe? What do dry cleaners really use to dry clean our clothes? Are there alternatives?

Are the aerosol hair sprays and deodorants we use safe for our environment? When we clean the bathroom tub and tile, are we releasing harmful chemicals into our water supply? Which, if any, of these things can be recycled? We will take a look at these and other concerns in this chapter.

DIAPERS

We spend $3.5 billion a year on the over 18 billion disposable diapers that we toss into our landfills. Disposable diapers alone account for anywhere between .5 percent to 2 percent of total landfill space taken up. That's a lot of diapers, but nowhere near as big a pile of garbage as newspapers, with 40 percent of landfill space taken up.

It is estimated that 75,000 metric tons of plastic and 1,265,000 tons of wood pulp from trees are used every year to make disposable diapers.

It is natural that people are concerned about diapers as garbage, and they are looking for alternatives. The problem with disposable diapers is both the amount of landfill space they take up and the fact that they do not readily decompose. Instead, they stay around in the environment for decades.

Kimberly-Clark Corporation, makers of Huggies, and Procter & Gamble, makers of Luvs and Pampers, say that their new lines of super-absorbent, ultra-thin diapers, because they are so thin and absorbent, have already begun to cut down on the amount of landfill space taken up by diapers by 50 percent.

Furthermore, diapers might be recyclable. A Seattle company is collecting used diapers from over 400 families. The waste is being flushed down the toilet, the paper is being treated and used as fertilizer, and the plastic sent to a plastic recycler.

The Cloth Diaper Alternative

Although most baby boom babies were raised in cloth diapers, baby boomers as parents have switched to disposables. Diaper services used to be as common as milk delivery. Both have become an endangered species. But cloth diaper services do exist. Look in the telephone book under Diaper Services and chances are you will find one. If not, call the National Association of Diaper Services, 2017 Walnut Street, Philadelphia, PA 19103, (215) 569-3650 or (800) 462-6237, for a list of diaper services in your area.

The diaper association says that you will save money by using cloth diapers. They say disposables cost thirteen to thirty-one cents each, while a diaper service charges seven to eleven cents per diaper. Washing your own diapers can save you even more money.

Changing disposable and cloth diapers is exactly the same. You remove the diaper and toss it in a plastic bag or in a hard plastic container. With disposables, the used diapers go to the landfill. With diaper service, the used diapers are picked up by a deliveryman and replaced with fresh ones. You don't even have to lug or dispose of those huge disposable diaper boxes.

If you are going to wash your own cloth diapers, you will need about four to five dozen diapers and a minimum of six diaper covers—many of these come with Velcro strips so you don't even have to use safety pins. Take the soiled diaper and dunk it in the toilet to remove waste, toss in a presoak bin filled with water and borax, then wash in the washing machine. Be sure to use mild soaps because a baby's skin is very sensitive.

Biobottoms is a new breathable wool diaper cover, as opposed to the usual plastic. It, too, has Velcro fasteners and can be used about five times before it should be washed along with the dirty diapers. Biobottoms, Box 6009, 3820 Bodega Avenue, Petaluma, CA 94953, (707) 778-7945, for mail order or the name of a store near you that carries them.

Bumkins is another solution to changing diapers. It is a washable cotton diaper wrapped in a nylon shell. You simply wash Bumkins and reuse them over and over again. Bumkins, 7720 East Redfield Road, Suite 4, Scottsdale, AZ 85260. (800) 553-9302.

THE LAUNDRY ROOM

Detergents. The biggest environmental problem associated with doing the laundry is the adverse effects phosphates can have on water quality. Phosphates are the chemical phosphorus that in agriculture is used as a fertilizer. It is a natural plant food. Too much phosphorus in the water causes algae blooms, which cause red and brown tides, which kill fish and shellfish.

The use of phosphates in laundry soaps is banned in many states, especially coastal states and areas like Long Island, New York, and areas around the Chesapeake Bay. The laundry soap you buy in many of these regions does not contain phosphates. The box will tell you if it contains phosphates or not. If you have any doubts, bear in mind that liquid laundry detergents do not contain any phosphates anywhere.

Bleach. Chlorine is a very dangerous and toxic chemical. But the amount used in liquid laundry bleach is safe to use if you follow the directions on the bottle. Overuse can cause water pollution. If you feel it is important for you not to use chlorine bleach, try dry oxygen bleaches. You can also cut the amount of chlorine bleach you use in half by adding one-quarter cup of baking soda to the wash and reducing the chlorine bleach by half.

DRY CLEANING AND SPOT REMOVERS

The solvents used in the dry cleaning process are hazardous chemicals and potential cancer causers. They include trichloroethylene, carbon tetrachloride, perchloroethylene, ethylene dichloride, naphtha, benzene, and toluene. Responsible dry cleaners use these products carefully, but accidents do occur and some of these dangerous chemicals are released into the atmosphere and water.

The best thing you can do is cut down on the amount of dry cleaning you send to the cleaners. Some items, like suits, are difficult to clean at home. But wool and cotton sweaters, silk blouses, and other fine washables can be washed at home. Just use a gentle soap like Woolite, set the machine on the gentle cycle, and wash.

Home dry cleaning fluids and spot removers should be treated like household hazardous waste. Use them up carefully or give them to someone who will. Do not throw in the trash. Save excess for household hazardous waste collection.

THE BATHROOM

Toilet paper. Toilet paper is not a serious environmental hazard. But the dyes and inks used in colored paper are mildly toxic and unnecessary. Use white toilet paper if you are concerned. But it does take a lot of trees to make all the toilet tissue we use. There are toilet papers made of recycled paper on the market. Check your local health food store or write Seventh Generation.

Feminine napkins. Avoid the use of plastic in tampons and other feminine hygiene products. Never toss plastic items in the toilet because they go directly into the water system and eventually into lakes, rivers, and streams. Place plastic and paper wrappers and tubes in the trash. Used cotton and paper products can be disposed of in the toilet.

Basin, tub, and tile cleaners. Most of these products are generally safe to use. The level of chlorine in scouring powders is very low, but if you want to use no chlorine, try Bon Ami. Most other bathroom cleaners are based on ammonia, again formulated at nonhazardous levels. But if you want to use no ammonia, try baking soda as a cleaner and either vinegar or lemon juice as a deodorizer.

COSMETICS AND PERSONAL CARE PRODUCTS

The aeosol cans used for hair spray and deodorant do not contain ozone-depleting chlorofluorocarbons, CFCs. After the public outcry against CFCs in aerosols, manufacturers switched to the use of methylene chloride as a propellant. But that chemical, too, was banned, in September 1989. The FDA considered methylene chloride to be cancer causing if inhaled. Many manufacturers will now be switching to a mixture of butane and alcohol in their spray cans.

If you are concerned about any chemicals in your spray cans, you may want to switch to pump-action sprays for hair care or roll-ons or sticks for deodorant.

Nail polish and nail polish remover contain chemicals that are hazardous. Use them up or give them to someone who will. Do not throw the bottles in the trash if they still contain liquids. Once at the landfill, they could leak out and seep, untreated, into the water system. Allow the nail polish to dry and then discard in the trash. Allow the polish remover to evaporate in the air and dispose of in trash.

Home permanent wave kits and hair relaxer kits usually contain sodium or potassium hydroxide, or thioglycolate. Unused portions should be flushed down the toilet with lots of water, not thrown in the trash.

Perfume, cologne, and after-shave contain large percentages of alcohol. Use them up or pour them down the drain with large amounts of water.

The contents of shampoo bottles are harmless to the environment. The bottles themselves are made of a type of plastic that is readily recyclable. Use up your shampoos, and then take the bottles to the recycling center.

7

HOME ENERGY CONSERVATION

One of the biggest worldwide sources of pollution is the energy we use to heat and cool our homes and power our appliances. It is also one that we can have a very dramatic impact on in reducing pollution levels.

Three of the fuels we use to generate energy, coal, fuel oil, and nuclear, are creating serious pollution problems. The burning of coal and fuel oil gives off carbon dioxide, sulfur dioxide, and other dangerous gases that have been linked to the greenhouse effect, to acid rain, and respiratory health problems for people.

The accidents at Three Mile Island and Chernobyl remind us that nuclear power stations can be risky ventures. Spent radioactive nuclear fuel is so dangerous that nobody is willing to dispose of it in their state. Cleaning up the handling of nuclear waste by the federal Department of Energy and Department of Defense is going to cost taxpayers millions of dollars over the next few years.

Natural gas, hydroelectric, and solar are much cleaner sources of power. But natural gas is restricted to certain parts of the country because of the extensive pipelines that have to be built to transport it. Furthermore, families that have fuel oil heaters find that because fuel oil is relatively cheap right now, it is not cost-effective to switch to natural gas.

Both solar and hydroelectric power were discovered by the public during the energy crisis of the 1970s. But since energy prices have declined in recent years, their cost effectiveness has declined, too. But of all the fuel sources, solar and hydro are the cleanest and the cheapest.

In this chapter, we are going to look at several things about home energy conservation. First we will discuss the health and environmental problems of acid rain, global warming, and smog caused by burning coal and fuel oil.

Next, we will take a tour of your house and do a home energy audit, and show you how you can cut your energy use with low-cost, no-cost techniques. Then, we will talk about the efficiency of your refrigerator, dishwasher, air conditioner, and other appliances.

We will explain how you can take advantage of existing solar power in your home for free and how landscaping can cut your fuel bills.

Finally, we will talk about water conservation as a way to cut energy costs, prevent pollution, and protect our resources. Let's get started.

ACID RAIN AND GLOBAL WARMING

Both of these are the result of burning coal and fuel oil, mainly in electric power plants, but also in schools, homes, and factories. (Trucks and automobiles do, too, but we talk about them in a different chapter.) Power plants and other industrial burners of fossil fuels have installed pollution control devices to cut down on emissions, and they have worked to some extent.

Acid rain, in particular, was caused when power plants built taller smokestacks, which were designed to eliminate pollution in areas around the power plants. The result was that the sulfur dioxide and nitrogen oxide gases went up farther into the atmosphere and were converted into acid rain. When acid rain falls, it kills the aquatic life in lakes and forests and crops on land. You can help cut down on acid rain by using less electricity and fuel oil in your home.

The same goes for global warming. The gases given off by burning oil and coal trap heat in the atmosphere, resulting in a gradual buildup of heat, which causes drought and perhaps a melting of the polar ice caps. Again, cut down on the electricity and fuel oil you use in your home. Here's how.

HOME ENERGY AUDIT

Before you spend any time or money on a do-it-yourself home energy audit, most electric companies and natural gas suppliers will do a free home energy audit on your house. Many will install a few inexpensive energy conservation measures for you as a service. My local utility, Niagara-Mohawk, did an energy audit on my house and sent me a computer printout of my energy problems and how much it would cost to fix each one, plus a payback time. Chances are you can get one, too. Look at your utility bill and call the customer service number listed. Tell them you want a free home energy audit. Simple.

Now, that doesn't mean you can't do an effective home energy audit for yourself. Why wait?

Before you start your home energy audit, I want to give you some basic principles to guide you and make this easy to understand. Simply put, your house breathes. By that I mean cold air and hot air are always trying to change places. You already know that hot air rises and cold air sinks. This hot-and-cold-air musical chairs goes on all the time.

Now, in the summer, when you have your air conditioner on, the cold air is always trying to get outside, and the hot air is always trying to get inside. The reverse is true in winter, when the hot air that you've spent so much money making hot is always trying to get out, and that shivering cold air outside wants to get in.

You basically want to stop this air flow. You do it by checking for leaks. The worst leakers are windows, even the window panes themselves, the attic, and doorways. Other important leakers are fireplaces, chimneys, baseboards, electric outlets, and where plumbing, heating, and wiring leave and enter the home.

The best time to check for leaks is on a windy day. Hold your hand near a suspected leak spot and feel for cool air. Caulking is the best way to stop leaks. Caulk on days when the temperature is above 40 degrees. There are several different types of caulk. Ask your hardware or paint store for advice on the best caulk in your area.

Windows and doors. The most energy-efficient windows are double or triple thermal panes with thin layers of air between the panes. But even these modern windows are leakers if not properly installed.

Check for cracked window panes, air leaks around panes, casings, and moldings, and caulk if necessary.

Install storm windows and storm doors. Caulk.

If you can't afford storm windows, try putting clear plastic sheets over windows. They are available at most hardware stores.

Drapes and shades can also help cut down on leaks. Pull them closed at night to keep warm air in. Open them during the day on windows where the sun shines in on the south side of the house. This is a form of passive solar heating.

A better solution is insulated shades or thermal shutters. Thermal shutters can be homemade of pillow material with a quilt covering, or rigid polystyrene board covered with fabric. Install these on the inside of the windows at night, and remove them during the day to allow sunlight to penetrate.

Check doors for leaks. Apply weather stripping if necessary. There are several types. Ask your hardware store for advice on the prices and durability of weather stripping in your area.

Fireplaces. Be sure the flue on your fireplace is closed. If you are not using your fireplace, it is best to close it off, rather than let all that expensive hot or cold air drift up the chimney.

Electrical outlets. Install insulating gaskets. If not in use, install plastic inserts in sockets.

Outside points of entry—for air, that is. Go outside and check for cracks around baseboards, sills, corners, and foundation. Caulk if necessary. Also check where wiring, plumbing, telephone wires, cable TV, and other lines enter your home. Be sure there is caulk at those points, too.

Insulation. Most older homes do not have any insulation either in the walls or in the attic. Many newer homes have inadequate insulation in walls and too often none in the attic.

Insulation is graded on its R-value. In general, the thicker the insulation, the greater the R-value, the greater capacity it has to insulate your house. R-19 is the recommended R-value for walls, R-30 for the attic. If you have a finished basement that you use as a living space, the R-7 is recommended for the walls.

If you are building a new house, or buying a new house,

be sure that the correct R-value is there. For an older house, attic insulation is the most cost-effective thing you can do. You can have it installed, or do it yourself; it is really very easy.

It is not a good idea to have your walls ripped apart to install wall insulation. Most people opt for blown insulation. If you live in a very cold climate, this might be a good idea. You can save money by doing it yourself. You can rent the blower, oftentimes the hardware store will loan it to you as long as you buy the insulation materials from them.

You simply drill a hole through the outside of your house between the studs, insert the nozzle, and blow in the insulation. The stuff comes in big bags. Repeat until all walls are filled. Plug up the holes, caulk, and paint.

With the exception of the blown-in insulation, most of the steps listed above have a payback period of less than three years. That means if you install them this year, they will have paid for themselves in saved energy cost in less than three years. Many energy companies will make low-cost loans to their customers for home energy conservation. Elderly and fixed-income people can usually get these things done for free from public service agencies.

THERMOSTAT SETTINGS

This is a very efficient and cost-effective way to reduce home energy consumption and costs. For every degree you set your thermostat back, you can save 1 to 3 percent of your heating and cooling costs.

But don't overdo it. Elderly people and people with health problems can get hypothermia, a lowering of body temperature and a slowing of body functions, if they get too cold. Don't set the temperature below 68 degrees in these cases.

Here's where you should set your thermostat:

65 to 70 degrees for winter daytime.
60 degrees for winter at night, after you've gone to
  bed.
55 degrees for winter when you are away from the
  house.
78 degrees for air conditioning in summer.

FURNACE, HOT WATER HEATER, AND AIR CONDITIONER EFFICIENCY

These three appliances use up most of the energy in your home. Naturally, if they are high-efficiency machines running at peak performance, you will use less energy, create less pollution, and save money.

Furnace. With electric heat, you don't have a furnace. You have baseboard heating elements that act much like little toasters. They turn red and give off heat.

Natural gas and fuel oil furnaces are similar in that they both burn fuel in an enclosed chamber. These chambers need to be cleaned every year and the jets cleaned. Also, be sure to change your furnace air filter every year. If your furnace is forced air, be sure that the blower is running smoothly and that the fan belts are tight and not cracked. Change your fuel filter if you have an oil furnace.

Hot water heaters. After heating the space in your home, heating the water for your home uses the most energy, 15.1 percent. The Missouri Department of Natural Resources has compiled a list of ten steps you can take to save energy use with your hot water heater.

1. Take quick showers, which uses less energy than long baths.

2. Fix leaky faucets. A dripping hot water faucet can waste as much as $3.50 in electricity each month.

3. Wrap your water heater in a blanket. Inexpensive, easy to install water heater insulation blankets can save the average family $17 a year.

4. Use flow-restrictor shower heads. You can cut water use by 50 percent. Less hot water used means less energy used.

5. Wait until there is a full load to run the dishwasher or clothes washing machine.

6. When purchasing a new water heater, buy the most energy efficient and heavily insulated model you can afford. You will save money in the long run.

7. Turn down the temperature on your water heater when you are away from home for more than seventy-two hours. A water heater will keep heating the water to a constant temperature, even if you aren't there.

8. Consider installing a solar hot water heater.

9. Insulate your hot water heater pipes.

10. Set the water temperature to 120°F or the lowest possible setting that you feel comfortable with.

Air conditioner. Uses 3.6 percent of home energy. Naturally, you want to buy an air conditioner with the highest efficiency rating possible. But before you buy a new one, make sure you are getting the most from your current model.

The key to cooling your house in summer is to keep the sun's harsh rays and heat out. Keep the blinds and windows closed and shut tight. When possible, keep your interior lights down low. You want to make your house cool, dark, and comfortable. Within reason. Don't create a dungeon. But you don't want the heat and humidity from the outside to seep in by opening the windows and shades on a hot summer day.

Circulate the cool air. One air conditioner plus a couple of strategically placed fans can cool off an entire house.

Don't let the sun beat down on your air conditioner parts outside. Create some shade with a trellis or awning.

Change the air conditioner filter as often as once a month during peak months.

Set the thermostat at 78 degrees. For every degree higher than that, you can save 4 percent of operating cost.

Leave the fan setting on manual so it will run continuously. The fan takes only one-fifth the energy to run as the compressor.

HOME APPLIANCES—REFRIGERATOR, DISHWASHER, WASHER/DRYER, STOVE

Refrigerator. Uses 5.7 percent of home energy. How big a refrigerator do I really need? One that is too big for your needs is just a waste of energy. A family of two needs an eight-cubic-foot refrigerator. Add one cubic foot for each additional person. Add two cubic feet if you entertain a lot. You need two cubic feet per person of freezer space.

▶ An energy-saver switch can save you 10 percent of energy cost.

▶ Self-defrosting models use more energy than standard ones.

▶ Upright freezers use more energy than chest types.

Dishwashers. Amount of energy use is included in hot water. If you use all the energy-saving features on your machine, it can be more economical than washing dishes by hand.

▶ An air dry selector, which turns heat off during the dry cycle, can save up to 30 percent of energy.
▶ Use short cycle selector for lightly soiled dishes.

Washer/dryer. Get the most efficient model you can. Keep the lint filter clean in the dryer. The heat from the dryer can be vented into the basement and considered an additional heat source for your house.

Stove. Uses 5.7 percent of energy.
▶ If you buy a gas range, get one with an electronic ignition system. You can save up to 30 percent over continuous pilot lights.
▶ If you buy an electric range, get the self-cleaning option. You will use some additional energy when it is cleaning itself, but you will save in the long run because it is so highly insulated.
▶ Glass windows that allow you to look inside the oven are better because it keeps you from opening the door to peek inside.

LANDSCAPING TO CUT ENERGY COSTS AND POLLUTION

As we've said before, the burning of fossil fuels, mainly coal and oil to generate heat and electricity, gives off carbon dioxide gas. CO_2 is linked to acid rain, smog, and global warming.

We can cut down on the amount of CO_2 produced by cutting down on the amount of energy we use. Another way is to plant more trees. Trees, along with grass and other plants, eat CO_2 and in turn give off oxygen. A mature tree can gobbble up fifty pounds of CO_2 from the air each year. Not much when you consider that 6 billion tons of CO_2 are emitted each year from burning of fossil fuels.

But every little bit helps. The American Forestry Association has launched a new program called "Global Re-leaf," with the slogan "Plant a tree, cool the globe." For information, write The American Forestry Association, P.O. Box 2000, Washington, D.C. 20013.

But other types of landscaping can help cut energy costs, too.

---

1. A windbreak of trees on the side of your house can reduce heating costs in winter by blocking cold winter winds.

2. Large mature trees can shade a house in summer, blocking the hot summer sun, and cut down on the amount of air conditioning you need.

WATER CONSERVATION

Most of us think that clean water is an unlimited resource. And to a large extent it is. But it takes a lot of energy not only to pump that water to your house, but also to purify it, to heat it for hot water uses, to flush it out of your home, and to clean it again in a sewage treatment facility.

Cutting down on your water use can save energy and extend our water resources. Most people can cut down on water use by 20 to 40 percent with simple conservation techniques. Here's how.

---

1. Check for and fix any leaks. A leaking toilet can waste 200 gallons a day. A steady drip wastes 20 gallons a day. Check for leaks by turning off all the water in the house. Find your water meter. Watch the number for fifteen to twenty minutes. If it registers a change, even when all the water is off, you have a leak.

2. Only do the laundry and use the dishwasher when you have full loads.

---

3. If doing dishes by hand, fill a rinse bucket rather than leaving the water running.

4. Don't use your garbage disposal if you have one. It uses an enormous amount of water and contributes to strains on sewage treatment facilities.

5. Place bricks or plastic water jugs in your toilet tank to cut down on the amount of water in each flush.

6. Install a low-flow shower head.

7. Don't leave the water running while brushing your teeth.

Too many of us remember the energy crisis of fifteen years ago as a time of deprivation. Energy conservation in the 1970s meant we shivered in our homes and waited on long lines at the gas pumps.

But energy conservation doesn't have to be like that. It is a sensible thing to do. Remember, a dollar saved is a dollar earned? Energy conservation in the 1990s means you can breathe cleaner air and cut down on your home fuel bills while driving a nice car and living in a warm and cozy home.

Sounds good to me.

RECYCLING AWAY FROM HOME

Anybody can recycle at home. It's easy. You have your recycling bins set up on the back porch. You can control what packaging your food comes in when you buy from the grocery.

But what about when you are away from home, at school, the office, or on the job? Can you recycle when you're traveling, eating in restaurants, and staying in hotels? Absolutely. Here's how.

AT SCHOOL, THE OFFICE, AND ON THE JOB

Schools and offices create a tremendous amount of paper. Bags and bags of paper are incinerated or tossed in the trash each day. Here are some tips on how you can reduce your school's or office's use of paper.

1. Buy paper that has been manufactured from recycled paper. Ask your current paper supplier to sell you recycled paper or find one who will. It is on the market and the more we demand it the more available it will be. Two companies offering recycled paper are: The Recycled Paper Company, 185 Corey Road, Boston, MA 02146, (617) 277-9901, and Earth Care Paper, Inc., Box 3335, Dept. 99, Madison, WI 53704, (608) 256-5522.

2. Use both sides of every sheet of paper you use. And that includes copies.

3. Use scratch paper for internal office memos and rough drafts.

4. Recycle all the office or school paper you can. Here's a list of things that can and cannot be recycled:

Can: white bond paper, Xerox copies, white scratch paper, colored scratch paper, colored tissue copies, computer printout paper, ledger sheets, envelopes without plastic windows.

Cannot: carbon paper, stencils, sensitized copy, Ditto masters, magazines, glossy paper, glue-bound material, blueprint paper.

5. Create a system to recycle your paper. The corporate offices of Waste Management, Inc., in Oak Brook, Illinois, use a dual wastepaper basket system. One basket for recyclable paper, and one for discards. Each employee is responsible to empty their trash into larger bins, but janitors could do this for you. The separated papers are then recycled.

The 1,100 employees at this office complex were given a slide presentation before the program began in 1988. They now recycle eighteen tons of paper each month and they have reduced their office waste by 50 percent. This item will become a very significant savings when offices are required to recycle in the very near future.

For more information contact Waste Management Systems, 3003 Butterfield Road, Oak Brook, IL 60521. (708) 572-8800. Another source of help is Diversified Recycling Systems, a company that designs and builds office recycling systems. Contact them at 5606 North Highway 169, New Hope, MN 55428. (612) 536-6664.

Paper is only part of what can be recycled at the office or school or on the job.

6. Set up regular recycling bins at work. Plastic, glass, and aluminum beverage containers can be recycled at work, too.

7. Lug-a-Mug. Bring your own coffee cup or mug and water glass from home. Use it and wash it and use it again. Don't use plastic, Styrofoam, or paper cups. They all end up in the trash.

8. Brown bagging. Bringing your own lunch is a good idea, but don't let it become another source of trash. Don't wrap your sandwich in plastic or waxed paper. At least use aluminum foil. The best idea is to place your sandwich in a hard plastic container shaped like a sandwich. Tupperware and Rubbermaid both make them. Ditto for salads. Put them in a hard plastic container rather than a flimsy one that will be thrown away. Soups and beverages should be carried in thermos-type containers. Don't place all this stuff in a paper bag that will just be thrown away. Buy a lunch pail. As corny as this may sound, you may just become a fashion trend setter.

9. The cafeteria. Insist that your employee cafeteria serves food on washable plates, bowls, coffee cups, drinking glasses, and silverware. Paper and soft plastic plates are too often trashed. An exception to this is Plastics Again, a recycler of used polystyrene foam cups and plates. Contact them at (508) 840-1521, or write 24 Jytek Park, Leominster, MA 01453.

RESTAURANTS

The amount of trash generated by restaurants is a function of the type of restaurant. The more convenient and fast your food is, the more wrapped up it is, the more paper and plastic trash it is going to produce. Let's take restaurants by category, discover the amount and types of trash they create, and figure out ways to cut down.

Do it fancy. Expensive, white-tablecloth restaurants create the least amount of paper and plastic trash because none of the food is wrapped up when they serve it to you. ·You generally have washable, reusable cloth napkins and table-cloths here, as well as washable plates, glasses and silverware. The only way you can cut down on trash is to clean your plate. Don't overorder and not eat all your food. Remember, food wastes make up nearly 8 percent of landfill space. You could also ask the waiter if the restaurant recycles its empty wine and beverage bottles.

Family restaurants and diners. You usually get washable silverware, plates, and glasses here. Request them if you don't. Most places have them somewhere in the back. Ask for cloth napkins rather than paper.

Too many of these restaurants are hooked on single-serving packages. Single-serving cream for your coffee, sugar packets, jam and jelly in plastic containers, ketchup and mustard pouches, double-wrapped plastic straws. The list goes on and on. Simply request that the cream be served in a little pitcher, the sugar free-flowing in a bowl, etc. It is in their best interest because soon enough they will have to pay extra to throw all this packaging in the trash and have it hauled away.

The owner will tell you they have individual servings for public health reasons. Well, nobody got sick from spooning sugar out of a bowl before single-serving packages were invented.

Fast food. Here is where packaging is king. Too many things are double- and triple-wrapped and there is discarded paper and plastic everywhere.

CFCs are no longer used in foam plastic packaging for the food service industry. CFCs have been linked to global warming, the greenhouse effect, and causing holes in our protective ozone layer, but fast food containers do not contribute anymore.

McDonald's is the most consistent user of polystyrene foam plastic, but they are not alone. Furthermore, McDonald's has started a campaign to recycle foam plastic.

The only way to cut down on the amount of packaging at fast food restaurants is to eat inside; at least you eliminate the paper carrying bag. What you can do is encourage the restaurant to find suppliers of recycled paper for their paper bags and napkins. Ask the owner if they have a separate bin for recycling foam plastic.

TRAVELING BY CAR AND AIRPLANE

The Car

> **1.** Carry two plastic or paper litter bags. One for recyclables like beverage containers, the other for trash. Many gas stations and rest stops in states that have mandatory recycling laws will have separate recycling trash cans for you. Some do, some don't, but you can be prepared.
>
> **2.** Avoid all drive-through shopping, including fast food restaurants, banks, dry cleaners, etc. Sitting in line with your motor running spews carbon dioxide and other gases that contribute to air pollution. Park your car, turn off the engine, and walk in to do your shopping.

The Airplane

Tips here are much the same as for eating in restaurants. Most airline food is overly wrapped in paper and plastic. Tell them you don't want it that way and ask them if they recycle the plastic trays as well as the beverage containers. The wrappers for peanuts and potato chips cannot be recycled.

HOTELS AND MOTELS

The only thing you can do here is not use the individual containers of shampoo and other toiletries given to you as a sign of hospitality. You can buy tiny hard plastic containers that you can refill and carry with you. That's what everybody did before there were freebies. Remember the plastic soap containers and toothbrush holders?

The bottom line for successful recycling away from home is to just pretend that you are at home. You wouldn't want people to walk into your home and dump trash in your living room, right? Just think of the whole world as your living room and keep it clean.

9 ▶ RESOURCES FOR RECYCLING

NATIONAL RESOURCES

United States Environmental Protection Agency, 401 M Street, SW, Washington, D.C. 20460. (202) 382-2090.

EPA Region 1, John F. Kennedy Federal Building, Boston, MA 02203. (617) 565-3400.

EPA Region 2, 26 Federal Plaza, New York, NY 10278. (212) 264-2525.

EPA Region 3, 841 Chestnut Street, Philadelphia, PA 19107. (215) 597-9814.

EPA Region 4, 345 Cortland Street NE, Atlanta, GA 30365. (404) 257-4727.

EPA Region 5, 230 South Dearborn, Chicago, IL 60604. (312) 353-2000.

EPA Region 6, 1445 Rose Avenue, Dallas, TX 75202. (214) 655-6444.

EPA Region 7, 726 Minnesota Avenue, Kansas City, KS 66101. (913) 236-2800.

EPA Region 8, 999 18th Street, Suite 500, Denver, CO 80202. (303) 293-1603.

EPA Region 9, 215 Fremont Street, San Francisco, CA 94104. (415) 974-8073.

EPA Region 10, 1200 Sixth Avenue, Seattle, WA 98101. (206) 442-5810.

Acid Rain Information Clearinghouse, 33 South Washington Street, Rochester, NY 14608. (716) 546-3796.

The Aluminum Association, 900 19th Street, NW, Washington, D.C. 20006. (202) 862-5163 or (202) 862-5162.

America The Beautiful Fund, 219 Shoreham Building, Washington, D.C. 20005. (202) 638-1649.

American Chemical Society, 1155 16th Street, NW, Washington, D.C. 20036. (202) 872-4600.

The American Forestry Association, 1516 P Street, NW, Washington, D.C. 20005. (202) 667-3300.

American Paper Institute, 260 Madison Avenue, New York, NY 10016. (212) 340-0600.

American Petroleum Institute, 1220 L Street, NW, Washington, D.C. 20005. (202) 682-8000.

The Center for Environmental Education, 1725 DeSales Street, NW, Washington, D.C. 20036. (202) 429-5609. Conservation of endangered species and their marine habitat.

Citizens Clearinghouse for Hazardous Waste, Box 926, Arlington, VA 22216. (703) 276-7070.

Clean Water Action Project, 317 Pennsylvania Avenue, SE, Washington, D.C. 20003. (202) 547-1196.

The Conservation Foundation, 1250 24th Street, NW, Washington, D.C. 20037. (202) 293-4800.

Council for Solid Waste Solutions, 1275 K Street, NW, Suite 400, Washington, D.C. 20005. (202) 371-5319. Plastics industry.

Defenders of Wildlife, 1244 19th Street, NW, Washington, D.C. 20036. (202) 659-9510.

Earth FIRST! Box 5871, Tucson, AZ 85703. (607) 622-1371. Radical activist environmentalists.

Earth Island Institute, 300 Broadway, Suite 28, San Francisco, CA 94133. (415) 788-3666. Rain forest and marine life preservation.

The Ecological Society of America, 730 11th Street, NW, Suite 400, Washington, D.C. 20001. (202) 393-5566. Society of professional ecologists, mostly professors.

Environmental Action, 1525 New Hampshire Avenue, NW, Washington, D.C. 20036. (202) 745-4870.

Environmental Defense Fund, 257 Park Avenue South, New York, NY 10010. (212) 505-2100. Founded by Rachel Carson, author of *Silent Spring.*

Friends of the Earth, 530 7th Street, SE, Washington, D.C. 20003. (202) 543-4312. Many local chapters.

General Federation of Women's Clubs, 1734 N Street, NW, Washington, D.C. 20036. (202) 347-3168. Very active in recycling projects.

Glass Packaging Institute, 1801 K Street, NW, Washington, D.C. 20006. (202) 887-4850.

Global Greenhouse Network, 1130 17th Street, NW, Washington, D.C. 20036. (202) 466-2823. Information on global warming.

Green Committee of Correspondence, Box 30208, Kansas City, MO 64112. (816) 931-9366. American Green political party based on European Green parties.

Greenpeace, 1432 U Street, NW, Washington, D.C. 20009. (202) 462-1177.

Institute of Scrap Recycling Industries, 1627 K Street, NW, Washington, D.C. 20006. (202) 466-4050. Scrap dealers.

Keep America Beautiful, 9 West Broad Street, Stamford, CT 06902. (203) 323-8987. Antilitter activists.

League of Women Voters, 1730 M Street, NW, Washington, D.C. 20036. (202) 429-1965. Very active in recycling. Call your local office first.

National Audubon Society, 950 Third Avenue, New York, NY 10022. (212) 832-3200.

National Coalition Against the Misuse of Pesticides, 530 7th Street, SE, Washington, D.C. 20003. (202) 543-5450.

National Recycling Coalition, 1101 30th Street, NW, Suite 304, Washington, D.C. 20007. (202) 625-6406

National Solid Waste Management Association, 1730 Rhode Island Avenue, NW, Washington, D.C. 20036. (202) 659-4613. Landfill and incineration industry.

National Wildlife Federation, 1400 16th Street, NW, Washington, D.C. 20036. (202) 797-6800.

Natural Resources Defense Council, 122 East 42nd Street, New York, NY 10168. (212) 949-0049. Released information on alar on apples.

Nature Conservancy, 1815 Lynn Street, Arlington, VA 22209. (703) 841-8737. Buys endangered land areas.

Rocky Mountain Institute, 1739 Snowmass Creek Road, Old Snowmass, CO 81654. (303) 927-3128. Energy conservation and solar experts.

Sierra Club, 730 Polk Street, San Francisco, CA 94109. (415) 776-2211.

Society of the Plastics Industry, 1275 K Street, NW, Washington, D.C. 20005. (202) 371-5200.

STATE RESOURCES

ALABAMA

The Alabama Conservancy, 2717 7th Avenue South, Suite 201, Birmingham, AL 35233. (205) 322-3126. Citizens' group.

Department of Environmental Management, 1751 Congressman W. L. Dickinson Drive, Montgomery, AL 36130. (205) 271-7700. Recycling, solid waste, air and water quality, hazardous waste.

ALASKA

Alaska Center for the Environment, 400 H Street, Suite 4, Anchorage, AK 99501. (907) 274-3621. Citizens' advocacy.

Alaska Environmental Lobby, P.O. Box 22151, Juneau, AK 99802. (907) 586-2345. Environmental lobbyists.

Department of Environmental Conservation, P.O. Box O, Juneau, AK 99811. (907) 465-2600. Recycling, solid waste, air and water quality, hazardous waste.

ARIZONA

Commission on the Arizona Environment, 1645 West Jefferson, Suite 416, Phoenix, AZ 85007. (602) 255-2102. Advises governor on environmental issues.

Community Energy Planner, Energy Office, 1700 Washington Street, Phoenix, AZ 85017. (602) 542-3633. Recycling.

Department of Environmental Quality, 2005 North Central Avenue, Phoenix, AZ 85004. (602) 257-2300. Air and water quality, solid waste.,

ARKANSAS

Department of Pollution Control and Ecology, 8001 National Drive, Little Rock, AR 72219. (501) 562-7444. Recycling, solid waste, water and air quality, hazardous waste.

CALIFORNIA

California Air Resources Board, P.O. Box 2815, Sacramento, CA 95812. (916) 322-2990. Air quality.

California Natural Resources Federation, 2830 Tenth Street, Suite 4, Berkeley, CA 94710. (415) 848-2211. Citizens' umbrella group.

California Waste Management Board, 1020 9th Street, Suite 300, Sacramento, CA 95814. (916) 322-3330. Recycling and composting.

Coalition for Clean Air, 309 Santa Monica Boulevard, Suite 212, Santa Monica, CA 90401. (213) 451-0651. Citizens for clean air.

Division of Recycling, Department of Conservation, 1025 P Street, Sacramento, CA 95814. (916) 323-3743. Recycling.

Ecology Center, 1403 Addison Street, Berkeley, CA 94702. (415) 548-2220. Environmental information center.

Environmental Defense Center, 906 Garden Street, Suite 2, Santa Barbara, CA 93101. (805) 963-1622. Public interest environmental law firm.

Environmental-Energy Education, Department of Education, 721 Capitol Mall, Sacramento, CA 94244. (916) 324-7187.

Water Resources Control Board, 901 P Street, Sacramento, CA 95801. (916) 445-3993. Water quality.

COLORADO

Colorado Environmental Coalition, 777 Grant Street, Number 606, Denver, CO 80203. (303) 837-8701. Citizens' environmental group.

Department of Health and Environmental Protection, 4210 East 11th Avenue, Denver, CO 80220. (303) 331-4510. Recycling, air and water quality, hazardous waste.

Office of Energy Conservation, 112 East 14th Avenue, Denver, CO 80203. (303) 866-2507.

CONNECTICUT

Connecticut Fund for the Environment, 152 Temple Street, New Haven, CT 06501. (203) 787-0646. Powerful citizens' group.

Council on Environmental Quality, Room 239, 165 Capitol Avenue, Hartford, CT 06106. (203) 566-3510. Reports to governor on environmental issues.

Department of Environmental Protection, State Office Building, 165 Capitol Avenue, Hartford, CT 06106. (203) 566-5599. Recycling, composting, air and water quality, solid waste, hazardous waste.

E-P Education Services, 15 Brittany Court, Cheshire, CT 06401. (203) 271-2756. Provides environmental information to educators.

DELAWARE

Department of Natural Resources and Environmental Control, 89 Kings Highway, Dover, DE 19903. (302) 736-4794. Recycling, air and water quality, solid waste, hazardous waste, energy.

DISTRICT OF COLUMBIA

Department of Public Works, 2000 14th Street, NW, Washington, D.C. 20009. (202) 939-8115. Recycling.

FLORIDA

Department of Environmental Regulation, 2600 Blair Stone Road, Tallahassee, FL 32399. (904) 488-4805. Recycling, composting, air and water quality.

Environmental Information Center of the Florida Conservation Foundation, 1191 Orange Avenue, Winter Park, FL 32789. (40&) 644-5377. Large citizens' group.

Florida Defenders of the Environment, 1523 NW 4th Street, Gainesville, FL 32601. (904) 372-6965. Activist citizens' group.

GEORGIA

Department of Natural Resources, Floyd Towers East, 205 Butler Street, Atlanta, GA 30334. (404) 656-3530. Recycling, air and water quality, solid waste, hazardous waste, energy.

Georgia Environmental Council, P.O. Box 2388, Decatur, GA 30031. (404) 262-1967. Umbrella organization of environmental groups.

HAWAII

Department of Health, P.O. Box 3378, Honolulu, HI 96801. (808) 548-6410. Recycling.

Department of Natural Resources, Box 621, Honolulu, HI 96809. (808) 548-6550. Air and water quality, hazardous waste.

Environmental Center, Water Resource Research Center, University of Hawaii, 2550 Campus Road, Honolulu, HI 96822. (808) 948-7361. Research and information dissemination.

Life of the Land, 19 Niolopa Place, Honolulu, HI 96817. (808) 595-3903. Activist environmental group.

Office of Environmental Quality Control, 465 South King Street, Room 104, Honolulu, HI 96813. (808) 548-6915. Clearinghouse and reports to governor.

IDAHO

Department of Health and Welfare, Statehouse, Boise, ID 83720. (208) 334-5879. Recycling, air and water quality, solid waste, hazardous waste.

Idaho Environmental Council, P.O. Box 1708, Idaho Falls, ID 83403. (208) 336-4930. Citizens' group.

ILLINOIS

Department of Energy and Natural Resources, 325 West Adams Street, Room 300, Springfield, IL 62704. (217) 785-2009. Recycling, solid waste, energy.

Illinois Environmental Council, 313 West Cook Street, Springfield, IL 62704. (217) 544-5954. Activist citizens' lobbying group.

Illinois Environmental Protection Agency, 2200 Churchill Road, Springfield, IL 62706. (217) 782-3397. Air and water quality, hazardous waste.

INDIANA

Hoosier Environmental Council, P.O. Box 1145, Indianapolis, IN 46206. (317) 636-8282. Activist citizens' group.

Indiana Department of Environmental Management, 105 South Meridian Street, Indianapolis, IN 46206. (317) 232-8603. Recycling, air and water quality, solid and hazardous waste.

IOWA

Department of Natural Resources, Wallace Building, East Ninth and Grand Avenue, Des Moines, IA 50319. (515) 281-5145. Air and water quality, recycling, solid and hazardous waste, energy.

Iowa Conservation Education Council, Route 1, Box 53, Guthrie Center, IA 50115. (515) 747-8383. Citizens' group.

Iowa Natural Heritage Foundation, Insurance Exchange Building, Suite 1005, 505 Fifth Avenue, Des Moines, IA 50309. (515) 288-1846. Natural resources protection group.

KANSAS

State Department of Health and Environment, Forbes Field, Building 740, Topeka, KS 66620. (913) 296-1594. Recycling, air and water quality, solid and hazardous waste.

KENTUCKY

Department for Environmental Protection, 18 Reilly Road, Frankfort, KY 40601. (502) 564-2150. Recycling, air and water quality, solid and hazardous waste.

Environmental Quality Commission, 18 Reilly Road, Ash Annex, Frankfort, KY 40601. (502) 564-2150. Advises governor on environmental issues.

Kentucky Association for Environmental Education, Blackacre Natural Preserve, 3200 Tucker Station Road, Jeffersontown, KY 40299. (502) 456-3747.

LOUISIANA

American Lung Association of Louisiana, 333 St. Charles Avenue, Suite 500, New Orleans, LA 70130. (504) 523-LUNG. Air pollution control group.

Department of Environmental Quality, Baton Rouge, LA 70804. (502) 342-1216. Recycling, composting, air and water quality, solid and hazardous waste.

MAINE

Department of Economic and Community Development, State House Station Number 130, Augusta, ME 04333. (207) 289-6800. Recycling, solid waste.

Maine Environmental Education Association, Chewonki Foundation, Wiscasset, ME 04578. (207) 882-7323.

Natural Resources Council of Maine, 271 State Street, Augusta, ME 04330. (207) 622-3101.

MARYLAND

Chesapeake Bay Foundation, 162 Prince George Street, Annapolis, MD 21401. (301) 268-8816. Works to protect Chesapeake Bay.

Department of the Environment, 2500 Broening Highway, Baltimore, MD 21224. (301) 631-3000. Recycling, air and water quality, solid and hazardous waste.

Maryland Environmental Trust, 118 North Howard Street, Suite 700, Baltimore, MD 21201. (301) 333-6440. Sponsors community recycling.

MASSACHUSETTS

Department of Environmental Quality Engineering, One Winter Street, Boston MA 02108. (617) 292-5856. Recycling and composting, air and water quality, solid and hazardous waste.

Environmental Lobby of Massachusetts, Three Joy Street, Boston, MA 02108. (617) 742-2553.

MICHIGAN

Department of Natural Resources, Box 30028, Lansing, MI 48909. (517) 373-1220. Recycling and composting, air and water quality, solid and hazardous waste.

West Michigan Environmental Action Council, 1432 Wealthy SE, Grand Rapids, MI 49506. (616) 451-3051. Citizen activists.

MINNESOTA

Pollution Control Agency, 520 Lafayette Road, St. Paul, MN 55155. (612) 296-6300. Recycling, composting, air and water quality, solid and hazardous waste.

MISSISSIPPI

Bureau of Pollution Control, Department of Natural Resources, P.O. Box 10385, Jackson, MS 39209. (601) 961-5171. Recycling, air and water quality, solid and hazardous waste.

MISSOURI

Department of Natural Resources, P.O. Box 176, Jefferson City, MO 65102. (314) 751-3332. In Missouri, (800) 334-6946. Recycling, air and water quality, solid and hazardous wastes, energy conservation.

MONTANA

Department of Natural Resources, 1520 East Sixth Avenue, Helena, MT 59620. (406) 444-6699. Energy conservation and alternative energy information.

Environmental Quality Council, State Capitol, Helena, MT 59620. (406) 444-3742. Environmental research arm of state legislature.

Montana Environmental Information Center, P.O. Box 1184, Helena, MT 59624. (406) 443-2520.

Northern Rockies Action Group, 9 Placer Street, Helena, MT 59601. (406) 442-6615. Activist citizens.

State Department of Health and Environmental Sciences, Cogswell Building, Capitol Station, Helena, MT 59620. (406) 444-2544. Recycling, air and water quality, solid and hazardous waste.

NEBRASKA

Department of Environmental Control, State House Station, Box 98922, Lincoln, NE 68509. (402) 471-2186. Recycling, air and water quality, solid and hazardous waste.

NEVADA

Division of Environmental Protection, Department of Conservation and Natural Resources, Capitol Complex, Nye Building, 201 South Falls Street, Carson City, NV 89701. (702) 885-4670. Air and water quality, solid and hazardous waste.

Energy Programs, Office of Community Services, Capitol Complex, Carson City, NV 89710. (702) 885-4908. Recycling.

NEW HAMPSHIRE

Council on Resources and Development, Office of State Planning, 2½ Beacon Street, Concord, NH 03301. (603) 271-2155. Conducts studies, makes recommendations to governor.

Department of Environmental Services, Waste Management Division, 6 Hazen Drive, Concord, NH 03301. (603) 271-3503. Recycling, composting, air and water quality, solid and hazardous waste.

New Hampshire Natural Resources Forum, Sky Farm, Box 341, Charlestown, NH 03603. (603) 826-5865. Citizens' group.

SPACE: Statewide Program of Action to Conserve Our Environment, P.O. Box 392, Exeter, NH 03833. (603) 778-1220. Activists.

NEW JERSEY

Association of New Jersey Environmental Commissions, P.O. Box 157, Mendham, NJ 07945. (201) 539-7547. Information, research.

Department of Environmental Protection, 401 East State Street, Trenton, NJ 08625. (609) 292-2885. Recycling, composting, air and water quality, solid and hazardous waste.

NEW MEXICO

Energy, Minerals, and Natural Resources, Energy Conservation Division, 525 Camino de los Marquez, Santa Fe, NM 87503. (505) 827-5990. Energy conservation.

Environmental Improvement Division, P.O. Box 986, Santa Fe, NM 87504. (505) 827-2850. Air and water quality, solid and hazardous waste.

Health and Environment Department, P.O. Box 968, Santa Fe, NM 87504. (505) 827-2780. Recycling.

NEW YORK

Department of Environmental Conservation, 50 Wolf Road, Albany, NY 12233. (518) 457-3446. Recycling, composting, air and water quality, solid and hazardous waste.

Environmental Planning Lobby, 33 Central Avenue, Albany, NY 12210. (518) 462-5526. Influences environmental legislation.

Environmental Protection Bureau, Department of Law, State of New York, 120 Broadway, New York, NY 10271. (292) 341-2451. Attorney general's office that sues on behalf of citizens to protect environment.

Office of Energy Conservation and Environmental Planning, New York State Department of Public Service, 3 Empire State Plaza, Albany, NY 12223. (518) 474-1677.

NORTH CAROLINA

Conservation Council of North Carolina, 307 Granville Road, Chapel Hill, NC 27514. (919) 942-7935. Citizens' group.

Department of Human Services, Solid Waste Management Branch, P.O. Box 2091, Raleigh, NC 27601. (919) 733-0692. Recycling.

Department of Natural Resources and Community Development, P.O. Box 27687, Raleigh, NC 27611. (919) 733-4984. Air and water quality, hazardous waste.

NORTH DAKOTA

Department of Health, Bismarck, ND 58505. (701) 224-2372. Recycling, air and water quality, solid and hazardous waste.

OHIO

Department of Natural Resources, Fountain Square, Columbus, OH 43224. (614) 265-6886. Recycling, energy conservation.

Environmental Protection Agency, 1800 Watermark Drive, Columbus, OH 43226. (614) 644-3020. Air and water quality, solid and hazardous waste.

Ohio Alliance for the Environment, 445 King Avenue, Columbus, OH 43201. (614) 421-7819. Pro-environment industry group.

Ohio Environmental Council, Suite 300, 22 East Gay Street, Columbus, OH 43215. (614) 224-4900. Citizens' group.

OKLAHOMA

State Department of Health, 1000 Northeast Tenth, Oklahoma City, OK 73152. (405) 271-5600. Recycling, air and water quality, solid and hazardous waste.

OREGON

Department of Environmental Quality, 811 SW Sixth Avenue, Portland, OR 97204. (503) 229-5300. Recycling, air and water quality, solid and hazardous waste.

Environmental Education Project, School of Education, Portland State University, P.O. Box 751, Portland OR 97207. (503) 229-4721.

Oregon Environmental Council, 2637 SW Water Avenue, Portland, OR 97201. (503) 222-1963. Citizens' group.

Oregon Natural Resources Council, Yeon Building, Suite 1050, 522 Southwest Fifth Avenue, Portland, OR 97204. (503) 223-9001. Citizens' group.

Oregon Public Interest Research Group, 027 SW Arthur, Portland, OR 97214. (503) 222-9641. Activists.

PENNSLYVANIA

Department of Environmental Resources, Box 2063, Harrisburg, PA 17120. (717) 787-1323. Recycling, air and water quality, solid and hazardous waste.

Pennsylvania Energy Office, P.O. Box 8010, Harrisburg, PA 17105. (717) 783-9981. Energy conservation.

Pennsylvania Environmental Council, 225 South 15th Street, Philadelphia, PA 19102. (215) 735-0966. Citizens' group.

RHODE ISLAND

Department of Environmental Management, 9 Hayes Street, Providence, RI 02908. (401) 277-2774. Recycling, composting, air and water quality, solid and hazardous waste.

Save The Bay, 434 Smith Street, Providence, RI 02908. (401) 456-1363. Dedicated to protecting the Narragansett Bay.

SOUTH CAROLINA

Department of Health and Environmental Control, J. Marion Sims Building, 2600 Bull Street, Columbia, SC 29201. (803) 734-4880. Recycling, air and water quality, solid and hazardous waste.

South Carolina Energy Office, P.O. Box 11405, Columbia, SC 29211. (803) 734-1740. Energy conservation.

SOUTH DAKOTA

Alternative Energy Program, Office of Energy Policy, 217½ West Missouri, Pierre, SD 57501. Recycling, energy conservation.

Board of Minerals and Environment, Department of Water and Natural Resources, Joe Foss Building, Pierre, SD 57501. (605) 773-3153. Air and water quality, solid and hazardous waste.

South Dakota Resources Coalition, P.O. Box 7020, Brookings, SD 57007. (605) 594-3558. Citizens' group.

TENNESSEE

Department of Health and Environment, 4th Floor, Customs House, Nashville, TN 37219. (615) 741-3424. Recycling, solid and hazardous waste.

Energy, Environment and Resources Center, University of Tennessee, 327 South Stadium Hall, Knoxville, TN 37996. (615) 974-4251. Information clearinghouse.

Environmental Action Fund, P.O. Box 22421, Nashville, TN 37202. (615) 244-4994. Lobbyists.

Tennessee Environmental Council, 1719 West End Avenue, Suite 227, Nashville, TN 37203. (615) 321-5075. Citizens' group.

TEXAS

Department of Health, 100 West 49th Street, Austin, TX 78756. (512) 458-7111. Recycling, air and water quality, solid and hazardous waste.

Texas Committee on Natural Resources, 5934 Royal Lane, Suite 223, Dallas, TX 75230. (214) 368-1791. Activist citizens.

UTAH

Utah Department of Health, P.O. Box 16700, Salt Lake City, UT 84116. (801) 538-6111. Recycling, air and water quality, solid and hazardous waste.

Utah Energy Office, 355 West North Temple Road, 3 Triad Center, Suite 450, Salt Lake City, UT 84180. (801) 538-5428. Energy conservation.

VERMONT

Department of Environmental Conservation, Waterbury Complex, 1 South, Waterbury, VT 05677. (802) 244-8755. Recycling, air and water quality, solid and hazardous waste.

VIRGINIA

Division of Litter Control and Recycling, Department of Waste Management, James Monroe Building, 11th Floor, Richmond, VA 23219. (804) 225-2667. Recycling.

Conservation Council of Virginia, P.O. Box 106, Richmond, VA 23201. (804) 353-6776. Citizens' lobby group.

Council on the Environment, 903 Ninth Street Office Building, Richmond, VA 23219. (804) 786-4500. Air and water quality, hazardous waste.

Division of Energy, 2201 West Broad Street, Richmond, VA 23220. (804) 367-1310. Energy conservation.

Piedmont Environmental Council, P.O. Box 460, Warrenton, VA 22186. (703) 347-2334. Northern Virginia conservation group.

WASHINGTON

Department of Ecology, Olympia, WA 98504. (206) 459-6000. Recycling, composting, air and water quality, solid and hazardous waste, energy conservation.

The Mountaineers, 300 Third Avenue, West, Seattle, WA 98119. (206) 284-6310. Outdoorsmen.

Washington Environmental Council, 76 South Main, Seattle, WA 98104. (206) 623-1483. Activist citizens' group.

WEST VIRGINIA

Department of Natural Resources, 1800 Washington Street East, Charleston, WV 25305. (304) 348-2754. Recycling, air and water quality, solid and hazardous waste.

West Virginia Highlands Conservancy, 1206 Virginia Street East, Suite 201, Charleston, WV 25301. (304) 343-2767.

WISCONSIN

Citizen's Natural Resources Association of Wisconsin, 2033 Menominee Drive, Oshkosh, WI 54901.

Department of Natural Resources, Box 7921, Madison, WI 53707. (608) 266-2621. Recycling, composting, air and water quality, solid and hazardous waste, energy conservation.

WYOMING

Environmental Quality Department, 122 West 25th Street, 4th Floor, Herschler Building, Cheyenne, WY 82002. (307) 777-7937. Recycling, air and water quality, solid and hazardous waste.

About the Author

Laurence Sombke is a New York–based writer who has written for *USA Today, USA Weekend* (for which he recently wrote a special supplement entitled "Clean Up Your Own Backyard"), *Esquire, New York* magazine, *Family Circle,* and other publications. Sombke has produced and announced a program, "Energy Issues," heard on radio stations throughout the Midwest and also has been a newswriter for ABC Radio News in New York. He has an M.A. in journalism.

Additional copies of *The Solution to Pollution* may be ordered by sending a check for $7.95 (please add the following for postage and handling: $1.50 for the first copy, $.50 for each added copy) to:

MasterMedia Limited
215 Park Avenue South
Suite 1601
New York, NY 10003
(212) 260-5600

Laurence Sombke is available for speeches and workshops. Please contact MasterMedia's Speakers' Bureau for availability and fee arrangements. Call Tony Colao at (201) 359-1612.

Other MasterMedia Books

THE PREGNANCY AND MOTHERHOOD DIARY: Planning the First Year of Your Second Career, by Susan Schiffer Stautberg, is the first and only undated appointment diary that shows how to manage pregnancy and career. ($12.95 spiralbound)

CITIES OF OPPORTUNITY: Finding the Best Place to Work, Live and Prosper in the 1990's and Beyond, by Dr. John Tepper Marlin, explores the job and living options for the next decade and into the next century. This consumer guide and handbook, written by one of the world's experts on cities, selects and features forty-six American cities and metropolitan areas. ($13.95 paper, $24.95 cloth)

THE DOLLARS AND SENSE OF DIVORCE, by Dr. Judith Briles, is the first book to combine practical tips on overcoming the legal hurdles with planning before, during, and after divorce. ($10.95 paper)

OUT THE ORGANIZATION: How Fast Could You Find a New Job?, by Madeleine and Robert Swain, is written for the millions of Americans whose jobs are no longer safe, whose companies are not loyal, and who face futures of uncertainty. It gives advice on finding a new job or starting your own business. ($11.95 paper, $17.95 cloth)

AGING PARENTS AND YOU: A Complete Handbook to Help You Help Your Elders Maintain a Healthy, Productive and Independent Life, by Eugenia Anderson-Ellis and Marsha Dryan, is a complete guide to providing care to aging relatives. It gives practical advice and resources to the adults who are helping their elders lead productive and independent lives. ($9.95 paper)

CRITICISM IN YOUR LIFE: How to Give It, How to Take It, How to Make It Work for You, by Dr. Deborah Bright, offers practical advice, in an upbeat, readable, and realistic fashion, for turning criticism into control. Charts and diagrams guide the reader into managing criticism from bosses, spouses, relationships, children, friends, neighbors, and in-laws. ($17.95 cloth)

BEYOND SUCCESS: How Volunteer Service Can Help You Begin Making a Life Instead of Just a Living, by John F. Raynolds III and Eleanor Raynolds, C.B.E., is a unique how-to book targeted to business and professional people considering volunteer work, senior citizens who wish to fill leisure time meaningfully, and students trying out various career options. The book is filled with interviews with celebrities, CEOs, and average citizens who talk about the benefits of service work. ($9.95 paper, $19.95 cloth)

MANAGING IT ALL: Time-Saving Ideas for Career, Family Relationships, and Self, by Beverly Benz Treuille and Susan Schiffer Stautberg, is written for women who are juggling careers and families. Over two hundred career women (ranging from a TV anchorwoman to an investment banker) were interviewed. The book contains many humorous anecdotes on saving time and improving the quality of life for self and family. ($9.95 paper)

REAL LIFE 101: (Almost) Surviving Your First Year Out of College, by Susan Kleinman, supplies welcome advice to those facing "real life" for the first time, focusing on work, money, health, and how to deal with freedom and responsibility. ($9.95 paper)

YOUR HEALTHY BODY, YOUR HEALTHY LIFE: How to Take Control of Your Medical Destiny, by Donald B. Louria, M.D., provides precise advice and strategies that will help you to live a long and healthy life. Learn also about nutrition, exercise, vitamins, and medication, as well as how to control risk factors for major diseases. ($12.95 paper)

THE CONFIDENCE FACTOR: How Self-Esteem Can Change Your Life, by Judith Briles, is based on a nationwide survey of 6,000 men and women. Briles explores why women so often feel a lack of self-confidence and have a poor opinion of themselves. She offers step-by-step advice on becoming the person you want to be. ($18.95 cloth)